Coating Substrates and Textiles

Andreas Giessmann

Coating Substrates and Textiles

A Practical Guide to Coating and Laminating Technologies

 Springer

Dr. Ing. Andreas Giessmann
Coatema Coating Machinery GmbH
Dormagen
Germany

Translation of the German 2nd edition published by Springer titled "Substrat- und Textilbeschichtung: Praxiswissen für Beschichtungs- und Kaschiertechnologien", ISBN 978-3-642-01416-1

ISBN 978-3-642-44114-1   ISBN 978-3-642-29160-9 (eBook)
DOI 10.1007/978-3-642-29160-9
Springer Heidelberg Dordrecht London New York

Printed on acid-free paper

Springer is part of Springer Science+Business Media (www.springer.com)

Preface to the 2nd German Edition

Despite rising expectations of materials and products, their life cycles or yield potential are constantly declining. Global competition is a strong factor in classic markets, and the years ahead will be heavily affected by the prevailing current financial crisis. In view of the banking system's problems, the automotive industry's heavy decline and the downward spiral of a large proportion of the economy, innovations and special products are in demand more than ever. Only companies entering the market with new innovations which have profit potential will survive.

Viewed broadly, a vertical diversification is taking place in the added value chain which should lead to the formation of new industry branches. Many companies are trying to improve or completely change the attributes of their products, to ultimately achieve added value. The plastics industry, too, has new coating and laminating methods which are described in the 2nd edition of this practical book.

In addition to the improvement of existing products, the development of new products is also enormously important. This is the only way to counter the effects of the general recession. Great potential for expansion remains, particularly in the field of renewable energies as well as that of thin coatings for electronics applications. However, new developments occur continually in conventional markets, too, such as technical textiles and medical technology, which are described in the new edition of this book. The subject areas were supplemented as follows:

Section 3.2.6: Slot die and pouring technologies
Co-author: Andrea Glawe

Section 3.5.1.5: UV curing

Section 3.5.1.6: Electron beam curing
Co-author: Karin Timmermanns

Section 6.2: Medical and hygiene products
Co-author: Moritz Graf zu Eulenburg

Section 6.3: Prepreg
Co-author: Regina Reuscher

Section 6.4: Thin film coating for the field of electronics
Co-author: Christoph Dittrich

Also new is Chap. 7 which deals with common chemicals such as PVC, PU and acrylates, as well as diverse additives.

On the subject of systems, there is a trend towards higher production speed and finishing services, while continuing to meet quality requirements. More flexible production with easier operation, reduced production costs and energy savings also contribute to improvements in terms of technical competitiveness.

Dormagen, June 2009   Andreas Giessmann

Preface to the 1st German Edition

The coating of substrates made of textiles, paper or film applies to the original material the characteristics that satisfy the rising quality demands the industry makes on the materials used. For this reason, the coating industry has experienced strong growth for years. This book strives to present a holistic view in order to do justice to the complexity of coating and to the significance of the coating industry. However, the focus is on the coating of textiles, which is a component of textile finishing. Here, no new material is created, but the substrate is given an additional function instead. Methods in the field of classic textile finishing include the dressing or the dying of textiles, but these will not be dealt with here.

When coating a flexible sheet of material (i.e. through continuous application of one or more layers in a liquid or paste form), the usual goal is indeed the generation of new materials and thus new areas of application. As a result, the coating of textiles can be categorised in an area similar to finishing. This is not only of elementary importance for the sake of defining terms, but also underlines the immense importance of textile coating in the context of the textile industry, or respectively, coating in its entirety.

In order to create new products, in addition to the actual coating with all the associated machinery, chemicals and processes, the extended environment must also be considered. The fast-moving market of today, as well as the financial pressure prevalent in companies (topics such as Rating and Basel II) forces the coating industry to take an extremely economical and ecological approach. That is why this book devotes particular attention to s the following topics: structure of a modern plant, processes and methods of coating, environmental protection, and economical aspects (investment).

Dormagen, November 2002   Andreas Giessmann

Contents

1 The Coating Industry 1
  1.1 Definitions 1
      1.1.1 Classification and Delineation of Coating 2
      1.1.2 Textile Coating 2
      1.1.3 Intelligent Textiles 3
  1.2 Historical Outline 5
  1.3 Economic Activity 5

2 Demands of a Coating Facility 11
  2.1 Logistic Integration Within a Coating Facility 12
      2.1.1 Technical Performance/Services 13
      2.1.2 Logistic Performance/Services 17
      2.1.3 Sales Performance/Services............................ 18
  2.2 Realisation of Logistical Integration 19

3 Basic Elements of Coating Systems 23
  3.1 The Roller.. 23
  3.2 Application Equipment 27
      3.2.1 Knife Systems 28
      3.2.2 Roller Application Systems 36
      3.2.3 Powder Coating....................................... 41
      3.2.4 Hot-Melt Method 41
      3.2.5 Injecting... 45
      3.2.6 Slot Die and Pouring Technologies..................... 47
      3.2.7 Dipping and Impregnating 49
      3.2.8 Point and Double-Point Coating 50
      3.2.9 Foam Coating 50
      3.2.10 Other Methods 51
  3.3 Material Transport 53
      3.3.1 Winding 53
      3.3.2 Guiding Web Movement 55
      3.3.3 Material Reserves (Compensator)....................... 56

        3.3.4   Accumulator Trough 56

        3.3.5   Tenter Frames 57

        3.3.6   Drive Technology 58

  3.4   Auxiliary Units 59

  3.5   Drying Systems 61

        3.5.1   Drying 61

        3.5.2   Minimum Exhaust Volume Flow 70

        3.5.3   Temperature Control Unit 70

        3.5.4   Thermal Oil Boiler 70

        3.5.5   Cooling 71

  3.6   Recording Data on Coating Systems for Quality Assurance
       Purposes 71

        3.6.1   Recording Measured Values 71

        3.6.2   Tension 73

        3.6.3   Temperatures 73

        3.6.4   Flow Conditions 74

        3.6.5   Coating Weight 75

        3.6.6   Material Moisture Level 75

4   Production Methods 77

  4.1   Coating Methods 78

        4.1.1   Transfer Coating 78

        4.1.2   Direct Coating 79

  4.2   Printing 80

  4.3   Lacquering 81

  4.4   Embossing 81

  4.5   Tumbling 82

  4.6   Combined Methods 83

  4.7   Flock Coating 84

  4.8   Laminating 85

        4.8.1   Dry Laminating 87

        4.8.2   Wet Laminating 88

        4.8.3   Flame Laminating 88

  4.9   Final Assembly and Inspection 88

  4.10 Reference Formulae 90

5   Paste Preparation 93

  5.1   Vacuum Sinus Dissolver 93

  5.2   Vacuum Filter 94

  5.3   Three-Roller Mill 95

  5.4   Wall-Mounted High-Speed Mixer 95

  5.5   Solvent Mixer 96

  5.6   Plastisol 96

        5.6.1   Preparations 96

        5.6.2   Using a High Speed Mixer 97

        5.6.3   Low Speed Mixers 98
        5.6.4   De-Airing... 99
        5.6.5   Filtration 99

6   Substrate Coating 101
   6.1   Properties of the End Product................................. 101
        6.1.1   Carrier Material....................................... 102
        6.1.2   Coating of the Surface 108
        6.1.3   Coated Textiles as Laminate Materials 108
   6.2   Medical and Hygiene Products 116
        6.2.1   Product Description 116
        6.2.2   Substrates and Chemistry.............................. 118
        6.2.3   Process Engineering 119
   6.3   Prepreg 121
        6.3.1   Product Description 121
        6.3.2   Substrates and Chemistry.............................. 122
        6.3.3   Process Engineering 124
   6.4   Thin Film Coating (In the Field of Electronics) 126
        6.4.1   Substrate Pre-Treatment............................... 127
        6.4.2   Coating 128
        6.4.3   Activation 130
        6.4.4   Substrate Post-Treatment (and Subsequent Processes) 132

7   Characteristics and Applications of Plastisols and Additives 135
   7.1   Introduction 135
        7.1.1   History 135
        7.1.2   PVC Resins for Plastisol 135
        7.1.3   What Is a Plastisol?................................... 136
   7.2   Rheology 137
        7.2.1   Viscosity According to Newton 137
        7.2.2   Shear Stress $\tau = F/S$ is in dyn/cm 138
        7.2.3   Non-Newtonian Fluids 138
        7.2.4   Rheologic Curves 139
        7.2.5   A Look at Actual Practice 139
        7.2.6   Thixotropy and Curing................................ 140
        7.2.7   Elements in the Composition of a Plastisol 140
        7.2.8   PVC Resins for Plastisols 141
   7.3   The Plasticisers 144
        7.3.1   Function... 144
        7.3.2   What are Plasticisers Made of? 145
        7.3.3   Their Performance Characteristics 145
        7.3.4   Classification.. 147
   7.4   Stabilising Agents, Accelerators and Co-Stabilising Agents 152
        7.4.1   Background 153
        7.4.2   Mercaptide Group 153

7.4.3 Carboxyl Group 154
7.4.4 Barium/Zinc Stabilising Agents 154
7.4.5 Calcium/Zinc Stabilising Agents 154
7.4.6 Accelerators 154
7.4.7 Conclusions 155
7.4.8 Co-stabilising Agents 156
7.4.9 Epoxy Stabilising Agents or Plasticisers 156
7.4.10 Epoxidised Soybean Oil 156
7.4.11 Epoxidised Linseed Oil 156
7.4.12 Octyl Epoxy Tallate and Isooctyl Epoxy Tallate 157
7.5 Foaming or Expanding Agents 157
7.5.1 Chemical Foaming 157
7.5.2 PVC Resin 158
7.5.3 Plasticisers 158
7.5.4 Accelerators 159
7.5.5 Foaming Agents 159
7.5.6 Azodicarbonamide 160
7.5.7 Sulfonhydrazide 161
7.5.8 Hydrocerole BIF (Böhringer Ingelheim) 161
7.5.9 Practical Aspects 162
7.5.10 Mechanical Foams 163
7.5.11 Surface-active Substances 164
7.5.12 Mixers for Mechanical Foams 164
7.5.13 Physical Foams 165
7.6 Extenders 165
7.6.1 Carbonate Extenders 165
7.6.2 Precipitated Calcium Carbonates 167
7.6.3 Carbonate-free Extenders 167
7.6.4 Additives with an Extender Function 168
7.6.5 General Influences of Extenders 169
7.7 Rheologic Modifiers 170
7.7.1 The Thickening Agents 170
7.7.2 Solvents 172
7.7.3 Substances for Flame Retardation and Smoke Suppression .. 173
7.8 UV Light Stabilising Agents 181
7.8.1 Other Successes in Weather Resistance 182
7.8.2 Important UV Light Stabilisers 182
7.9 Adhesion Promoters for Plastisols 183
7.9.1 Single-Component Adhesives 183
7.9.2 Two-Component Adhesives 184
7.10 Biocides 184
7.10.1 Microbiological Attack 185
7.10.2 Fungi 185
7.10.3 Bacteria 186
7.10.4 Algae 186
7.10.5 Common Biocides for Plastisol 186

7.11 Other Raw Materials 187
    7.11.1 Antistatic Agents 187
    7.11.2 Calcium Oxide 187
    7.11.3 Air Releasing Agents 187
7.12 Economy 188
    7.12.1 Relative Density 188
    7.12.2 Formula Examples and Costs 188

8 Emission Control in Coating Companies 193
8.1 Air Pollution 193
    8.1.1 Emission 196
    8.1.2 Immissions 196
    8.1.3 Threshold Values 196
    8.1.4 Air Quality Control Regulations 197
8.2 Emission Control Procedure 198
    8.2.1 Procedure with Solid and Liquid Pollutants 198
    8.2.2 Procedure for Vaporous and Gaseous Pollutants 202
    8.2.3 Combination Procedures 211
8.3 Requirements for Emission Control Systems 211
8.4 Emission Control System Selection 212

9 Investment Profitability Analysis 217
9.1 Strategic Profitability Analysis 219
9.2 The Business Strategy 220
9.3 Production Strategy 221
9.4 Coating Plant Cost Elements 222
    9.4.1 Investment Spending 223
    9.4.2 Operating Costs 223
    9.4.3 Annuity 228
    9.4.4 Total Costs 229
9.5 Performance Elements in Coating Plants 229
    9.5.1 Flexibility 230
    9.5.2 Quality 230
    9.5.3 Quality 230
9.6 Operational Profitability Analysis 231
9.7 Planning Result Calculation 234
    9.7.1 Annual Sales 234
    9.7.2 Profits 235
    9.7.3 Summary of All Costs and Profits 235
    9.7.4 Comparison Figures 236

Bibliography 237

Index 241

Chapter 1
The Coating Industry

1.1 Definitions

When one or more synthetic layers are continuously applied in a liquid or paste form to a flexible sheet of material (substrate), a new material is created which combines the diverse characteristics of the joined materials (Werner 1986). As such, physical characteristics like tear propagation resistance and tensile load are largely influenced by the substrate. However, surface characteristics including resistance to solvents, colour-fastness, air permeability, water permeability, non-combustibility and many others are essentially determined by the coating. Depending on the kind and quantity of the components, as well as their bonding and the applied coating process, the characteristics of the composite material being created can be planned to match the future intended use (Enka 1987).

Coating and laminating are finishing processes for textiles, foils, paper and other substrates whose purpose is to create additional functionality and/or an appreciation in value of the materials, brought about by special characteristics (Smith 1999). Differentiation can be made between the following processes:

Coating Coating means the application of synthetic substances which, depending on the method of processing—as described more precisely later—can be applied to the carrier (textile sheets, paper, synthetic film) as paste, plastisol, organosol or melted mass (Fries 1992, p. 104).

Laminating A lamination is the joining of two carrier materials by adhesion. Here an additional joining element is used, namely the adhesive or the cohesion of two materials. Dispersions and organosols are both used as adhesives (Fries 1992). These, in turn, are applied (coated) using an application system, which will be described later.

Coating or laminating can be done for purposes of either aesthetic or functional optimization.

A. Giessmann, *Coating Substrates and Textiles*,
DOI 10.1007/978-3-642-29160-9_1, © Springer-Verlag Berlin Heidelberg 2012

1.1.1   Classification and Delineation of Coating

The methods available to coat sheet materials are manifold. Consequently, the commodities which result from substrate coating as end products are correspondingly varied. It is also possible to produce different kinds of the same end products.

The following example illustrates this clearly: in order to manufacture imitation leather either the *spread coating method* (direct or transfer method), the *calender method*, or the *coagulation method* can be used. Because of the variety of production technologies, different degrees of product quality are generated for diverse areas of application. Thus, when using the spread coating method in combination with a paste-like media, and when applied without tension, an end product can be manufactured that is very close to genuine leather in terms of feel, suppleness and appearance. Even the fashionable effects of genuine leather are apparent in the imitation leather.

In contrast, in the calender method the chemicals are formed into a film by means of pressure and heat, which is further processed into an end product, either with or without carrier material. This film contains stresses which can lead to premature brittleness and hardening of the end product during processing. For this reason, the areas of use of these films are different than those of products manufactured using the spread coating method. They are suitable, among other things, for vinyl flooring or tarpaulin materials.

Manufacture using the coagulation method is limited with respect to the substrate used. Since solvents are used in high concentrations, the manufacturer is responsible for proper handling and disposal. The result—a leather-like product with the feel of leather, formed on a spunbonded non-woven fabric—is skin-friendly and can be used in applications especially in the shoe, decoration and clothing industries.

Since the finishing of textiles is constantly increasing in importance, Fig. 1.1 provides an overview of the status of coating in the area of textile finishing.

1.1.2   Textile Coating

A range of very different areas of application exists for the coating of textile carrier materials:

- Agricultural textiles
- Construction textiles
- Clothing
- Geotextiles
- Home textiles
- Industrial textiles
- Medical textiles
- Mobile technology
- Ecological textiles
- Packaging materials

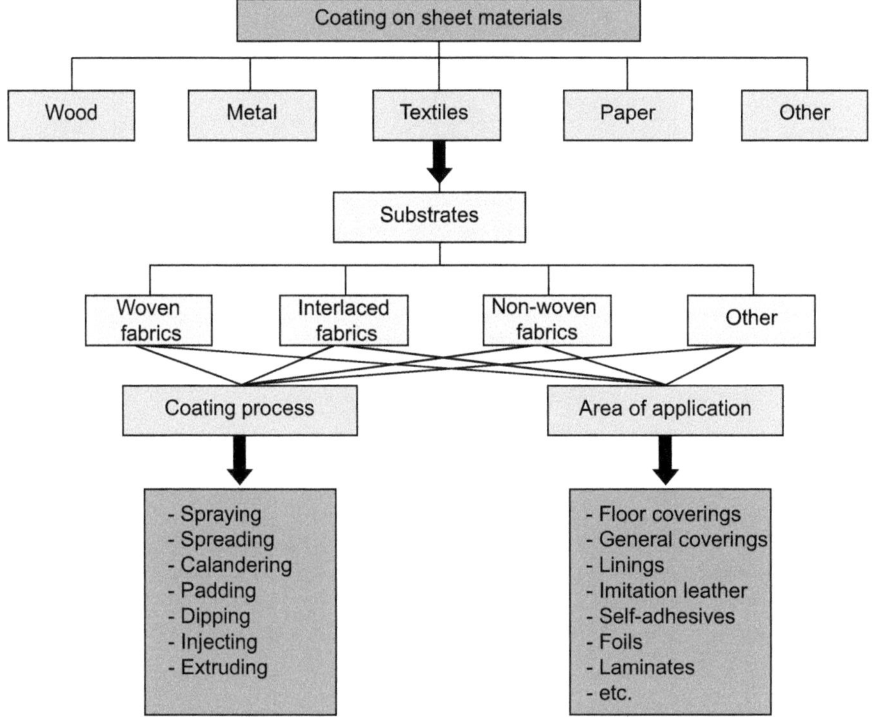

Fig. 1.1 Position of coating in textile finishing

- Protective clothing
- Sports clothing

1.1.3 Intelligent Textiles

The proportion of coated substrates is rapidly increasing; e.g. in coated textiles alone the rate of growth in recent years has reached double digits.

Coating gives the textile a functional surface. Substrates with a very specific, individually designed function, which can react actively to corresponding situations in the environment, are designated *Intelligent Textiles* or *Smart Textiles* (Fig. 1.2).

While technical textiles include those used for e.g. the drainage layer in road construction or to strengthen embankments and slopes, or as filter systems, in terms of intelligent textiles, possibilities range from the development to the implementation of sophisticated electronic sensors and data transfer systems which can serve to monitor patients, for example (Plank 2001).

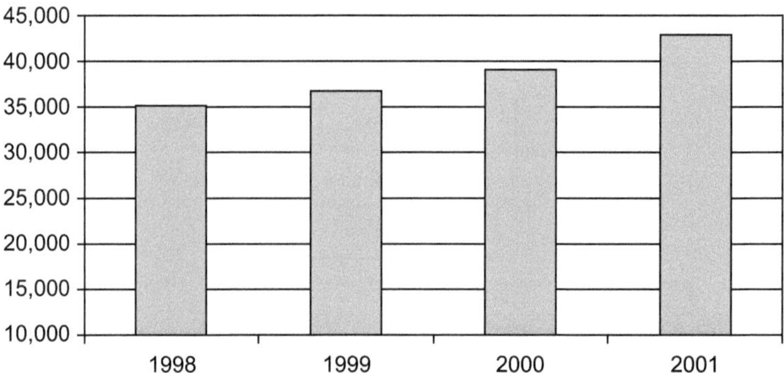

Fig. 1.2 Production of coated fabrics in thousands (statistical analysis of the TVI Verband)

Nanotechnology is another interesting field of application for the coating industry. By introducing or applying nano-particles to paper, film or textiles, it is possible to create self-cleaning surfaces, among other things. The intention of the so-called lotus effect is to ensure that dirt particles are directly washed away along with the flow of liquid, without any further treatment.

Furthermore, attempts are being made to introduce natural cells to a textile—particularly in non-woven fabrics—in order to generate biohybrid substitute organic materials. The Hohenstein Institute for international textile research has compiled the following classification for intelligent textiles (Mecheels 2001):

Transfer Systems Transfer of drugs or active agents from a coated or impregnated textile to the human skin, or absorption of substances from the skin into the textile.

Adaptive Systems The substrates automatically adapt to their surroundings; i.e. they react to light, heat, moisture and other influences from their surroundings.

Smart Clothing Clothing textiles in which electronic components have been integrated.

Transponder Systems Laser coding and radio frequency identification in textiles, for seamless and overarching information throughout the entire textile chain.

Micro and Nano-Systems Micro and nano-technologies integrated in textiles as self-acting control or steering and regulating mechanisms.

Opportunities for the development of intelligent textiles lie in the necessary interdisciplinary cooperation of different areas of technology. The networking of science disciplines such as electrical engineering, medicine, materials science, and information technology opens up new possibilities. Visions of the future reveal clothing in which mobile telephones, computers or health test systems for daily use are already integrated.

1.2   Historical Outline

The origins of coating are not clearly documented in literature. It is assumed that the first technical textiles were made in central Asia where winter housing, for example, was made out of dense felts, which were brushed or simply coated with mud or other natural materials (Reetz 2001). The first kinds of coatings existed already among the early Maya. They used the sap from rubber plants to make weatherproof (rainproof) clothing, by applying the sap to textiles and drying them over a fire (Sen 1999).

Attempts at coating can also be found in the case of ropes, cords and sails from this time. The materials were coated by primitive means, so that they would not become waterlogged.

The documented beginnings of coating occurred in 1783 in Paris as a balloon coated with natural rubber was made for filling with hydrogen.

In 1823, a Scot named Macintosh had the first rain clothing patented, which he produced by joining a rubber layer between two textiles. Attempts were made to imitate genuine leather by coating textiles with synthetic materials as early as 1830.

Thanks to the discovery of synthetic rubber, considerable developments of polymer substances were started in 1909 which, today, can be found in a wide range of high-tech products such as polyethylenedioxithiophene (PEDT = antistatic finish) or liquid crystal polymer (LCP for the coating or encasing of fibreglass).

In terms of geography, America was the first leader in coating. Later on, this industry established itself predominantly in Austria and then in Germany, where it has experienced a strong upsurge until today (Textilveredlung 1992). Coating of films and paper joined textile coating in 1950. In addition, thermoplastic materials were used in the 1970s.

1.3   Economic Activity

The economic climate index prepared by the Ifo Institute shows that the figures provided by companies in all areas worsened in the course of the year 2008 because of the general recession.

Thus, the textile sector reduced production in October 2008 in the two digit range (-10.8 %). Overall, the industry produced 12.8 % less goods in October and 5.9 % less in the course of the first 10 months in inland, in comparison with 2007 (Table 1.1).

The following studies and publications by the VDM, the Association of German Machinery Manufacturers (Verein deutscher Maschinenbauer), also clearly show that the decline at machine manufacturers in Germany alone exceeded 60 % in some sectors.

The figures not only confirm the recession in Germany, but document that, in this case, the proper term would be 'world economic crisis'. According to statements in the relevant sections of the press, this situation will continue at a minimum until and including 2010.

As a result, the innovative companies which are moving away from the crisis in an anti-cyclical manner and are now investing in the development and implementation of new products, become all the more important.

Table 1.1 Key figures for the textile and clothing trades

Year 2008	Textiles		Clothing		Textiles + clothing	
		$\pm$ % amount of previous year		$\pm$ % amount of previous year		$\pm$ % amount of previous year
1. Employees						
October	66,148	−4.0	32,039	−4.7	98,187	−4.2
January–October	66,937	−3.1	32,560	−5.0	99,498	−3.7
2. Gross pay and salary (in million €)						
October	169	−3.2	85	−1.0	254	−2.5
January–October	1,643	−1.6	808	−2.8	2,451	−2.0
3. Work hours performed (in 1,000)						
October	9,021	−5.1	4,064	−5.4	13,085	−5.2
January–October	87,574	−3.0	40,247	−5.4	127,821	−3.8
4. Sales (in million €)						
October	988	−8.3	669	−5.4	1,657	−7.1
January–October	9,543	−3.0	7,154	−3.1	16,697	−3.1
5. Production (index: 2,000 = 100)						
September	83.6	−0.7	44.4	−12.6	69.8	−3.7
October	82.6	−10.8	35.8	−20.6	66.1	−12.8
August–October	–	−7.7	–	−19.9	–	−10.8
January–October	–	−2.2	–	−17.5	–	−5.9
6. Incoming orders (index: 2,000 = 100)						
September	82.2	−2.8	79.1	−4.0	80.9	−3.3
October	83.0	−9.5	59.2	−7.1	73.0	−8.8
August–October	–	−8.1	–	−10.4	–	−9.2
January–October	–	−3.0	–	−5.6	–	−4.2

7. Incoming orders and production indices according to selected business areas, textile trade

Incoming orders (2,000 = 100)	Change from previous year (%)	Production (2,000 = 100)	Change from previous year (%)	Sales in million €	Change from previous year (%)
Preparation of spun fibres and spinning mills					
October	0.0 −100.0	0.0	−100.0	76	27.2
September	0.0 −100.0	59.8	−5.2	70	−24.3
January–October	39.4 −33.7	54.3	−18.8	749	−23.1
Weaving mill					
October	0.0 −100.0	0.0	−100.0	178	−14.2
September	0.0 −100.0	66.8	−4.0	175	−5.4
January–October	47.4 −27.0	58.3	−15.9	1,743	−7.1
Textile finishing					
October	0.0 −100.0	−0.0	−100.0	74	−1.0
September	0.0 −100.0	71.7	−4.8	72	10.2
January–October	54.7 −26.5	66.1	−12.8	735	7.1
Ready-made textile goods (without clothing)					
October	0.0 −100.0	0.0	−100.0	143	−1.3
September	0.0 −100.0	81.1	−4.5	159	12.7
January–October	73.3 −19.2	74.9	−12.5	1,355	0.5

Table 1.1 (Continued)

Misc. textile trades (without amount of knit fabrics)						
October	0.0	−100.0	0.0	−100.0	404	−4.3
September	0.0	−100.0	113.5	3.0	409	7.1
January–October	97.5	−17.3	100.5	−8.5	3,907	0.5
Non-woven fabrics and products						
October	0.0	−100.0	0.0	−100.0	115	0.4
September	0.0	−100.0	131.8	6.6	114	7.3
January–October	110.7	−17.0	115.2	−7.2	1,120	4.1
Interlaced and knit fabrics						
October	0.0	−100.0	0.0	−100.0	43	−12.9
September	0.0	−100.0	74.7	−1.3	43	−2.1
January–October	61.0	−24.6	65.3	−16.6	431	−5.1
Interlaced and knit end products						
October	0.0	−100.0	0.0	−100.0	71	−5.8
September	0.0	−100.0	55.3	−7.4	70	−0.1
January–October	57.3	−23.8	51.7	−12.2	623	0.9
Clothing trades						
Leather clothing						
October	0.0	−100.0	0.0	−100.0		
September	0.0	−100.0	103.4	−9.7		
January–October	79.6	−39.1	93.7	−5.7		
Clothing (without leather clothing)						
October	0.0	−100.0	0.0	−100.0	665	−5.6
September	0.0	−100.0	44.7	−12.5	878	2.4
January–October	72.3	−20.8	37.1	−24.9	7,109	−3.1
Work and professional clothing						
October	0.0	−100.0	0.0	−100.0	31	5.9
September	0.0	−100.0	100.4	8.0	29	11.4
January–October	88.6	−15.4	81.4	−12.1	263	8.6
Outer clothing (without work and professional clothing)						
October	0.0	−100.0	0.0	−100.0	484	−5.7
September	0.0	−100.0	41.6	−14.4	674	0.3
January–October	74.5	−20.5	34.2	−25.6	5,324	−4.6
Underwear						
October	0.0	−100.0	0.0	−100.0	106	−14.6
September	0.0	−100.0	37.6	−20.3	125	5.4
January–October	56.1	−27.5	32.6	−32.5	1,079	−5.3
Other clothing and accessories						
October	0.0	−100.0	0.0	−100.0	43	17.4
September	0.0	−100.0	54.3	−3.7	51	21.2
January–October	95.0	−9.9	45.8	−16.0	442	16.9

December 2008 issue
German exports of textile machines in million €

Selection of countries refers to rankings in the year 2007

Year	Textile machines all countries	China/HK	Turkey	India	Italy	USA	Czech Republic
2003	3,834	1,004	571	109	173	277	117
2004	3,605	1,015	393	142	216	256	91
2005	3,417	741	309	232	179	315	79
2006	3,621	929	237	368	172	192	92
2007	3,881	1,004	408	301	186	182	117
Quota in %							
2007		25.9	10.5	7.8	4.8	4.7	3.0
Jan - Sep. 2008		26.5	5.4	7.0	4.7	8.1	3.4
Change in %							
Sep. 2008/2007	-38.5	-60.7	-79.2	-39.4	-29.6	24.4	-19.1
Jan. - Sep. 2008/2007	-17.0	-13.8	-54.8	-26.2	-19.5	46.5	0.4
Jan. - Sep. 2008	2,399	636	129	167	113	195	81
Jan.	297	85	21	26	14	11	11
Feb.	309	81	19	18	14	23	10
Mar.	275	76	13	20	11	23	10
1st Quarter	882	241	53	65	40	56	31
April	358	114	18	19	18	30	16
May	226	54	12	19	14	16	7
June	273	65	16	21	12	20	8
2nd Quarter	857	233	46	59	44	66	31
July	248	75	11	14	11	19	5
Aug.	218	49	12	15	7	35	5
Sep.	194	38	7	14	11	19	9
3rd Quarter	660	162	30	43	29	73	19
Oct.							
Nov.							
Dec.							
4th Quarter							

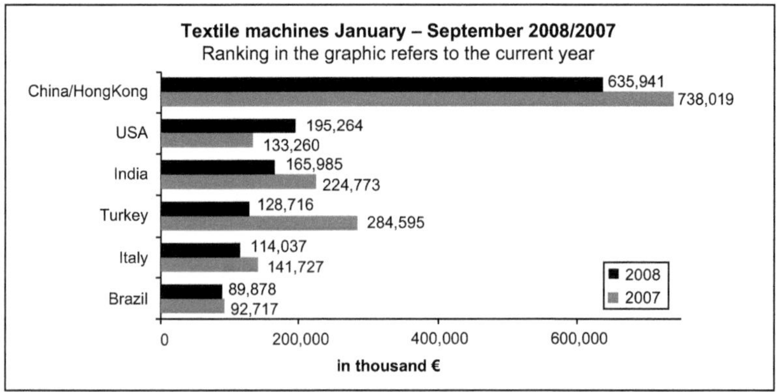

Textile machines

December 2008 issue
German exports of finishing machines in million €

Selection of countries refers to **rankings in the year 2007**

Year	Total	China/HK	Brazil	Russia	Turkey	USA	Italy
2003	635	107	7	8	77	48	40
2004	626	107	6	11	49	72	29
2005	776	75	62	4	71	97	27
2006	706	89	13	7	46	55	39
2007	684	72	43	41	40	40	37
Quota in %							
2007	17.6	10.5	6.3	6.0	5.8	5.8	5.4
Jan - Sep. 2008	21.7	13.1	5.8	2.9	5.8	9.2	5.2
Change in %							
Sep. 2008/2007	-10.3	12.3	-94.6	-74.1	-37.7	113.6	-53.7
Jan. - Sep. 2008/2007	0.7	12.7	-24.8	-39.2	-0.9	70.9	7.3
Jan. - Sep. 2008	520	68	30	15	30	48	27
Jan.	81	17	0,9	4	10	2	6
Feb.	67	7	8	1	3	3	3
Mar.	63	7	15	2	5	3	2
1st Quarter	211	31	23	7	18	8	11
April	63	10	1	3	3	3	4
May	51	4	2	0,6	2	3	3
June	53	8	0.9	1	2	4	3
2nd Quarter	167	22	4	5	7	10	10
July	35	2	0,6	2	2	4	2
Aug.	56	8	2	0,4	1	18	2
Sep.	51	5	0,4	1	2	8	2
3rd Quarter	142	15	3	3,4	5	30	6
Oct.							
Nov.							
Dec.							
4th Quarter							

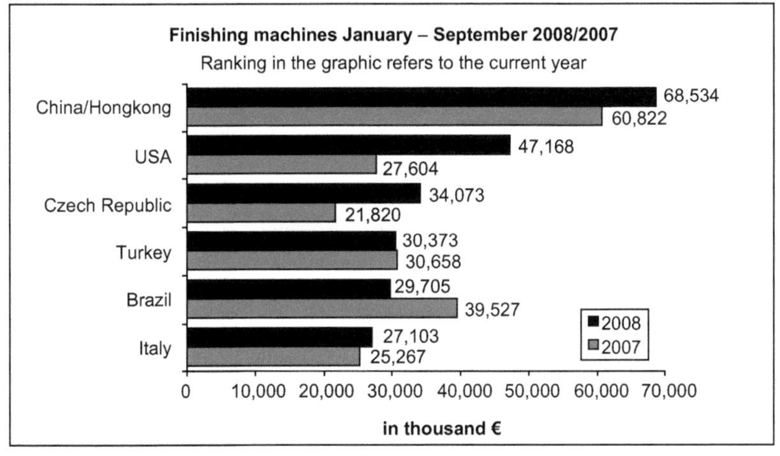

Finishing machines January – September 2008/2007

Ranking in the graphic refers to the current year

Textile machines

December 2008 issue
German exports of textile machines in million €

Year	Textile machines	Spinning mills	Weaving mills	Interlace and knitting machines	Textile finishing
2003	3,834	1,714	297	1,188	635
2004	3,605	1,522	363	1,094	626
2005	3,417	1,346	257	1,038	776
2006	3,620	1,352	344	1,217	706
2007	3,881	1,614	351	1,232	684
Quota in %					
2007		41.6	9.0	31.7	17.6
Jan. -Sep. 2008		36.5	8.4	33.4	21.7
Change in %					
Sep. 2008/2007	-38.5	-46.5	-62.3	-38.5	-10.3
Jan.-Sep. 2008/2007	-17.0	-24.5	-26.9	-14.6	-0.7
Jan.-Sep. 2008	2.399	875	202	803	520
Jan.	297	84	26	106	81
Feb.	309	114	30	98	67
Mar.	275	115	29	69	63
1st Quarter	882	312	85	273	211
April	358	132	30	134	63
May	226	83	20	73	51
June	273	113	24	83	53
2nd Quarter	857	328	74	290	167
July	248	83	16	114	35
August	218	87	17	58	56
September	194	65	10	68	51
3rd Quarter	660	235	43	240	142
October					
November					
December					
4th Quarter					

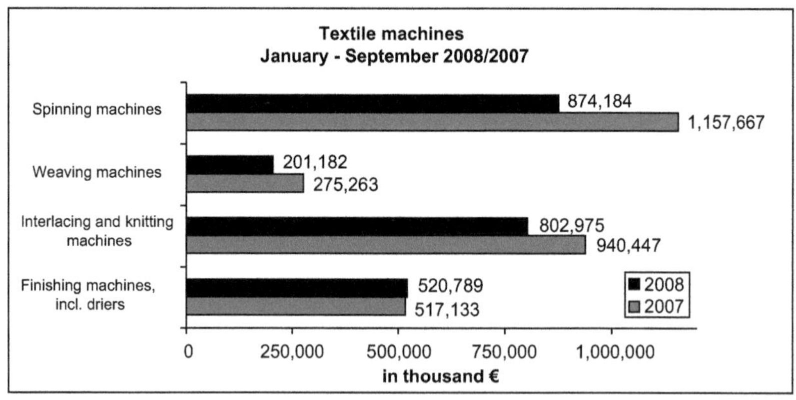

Chapter 2
Demands of a Coating Facility

When planning a coating facility, many considerations must be taken into account, because the success of a company is the result of many different factors. Included here are the location and layout of the plant, the chemicals which will be used and their preparation, and the investment in specific, additional components.

Plant Planning and Layout of the Plant The size of the individual sections such as raw materials storage, paste preparation room, production facilities or end product storage is mainly dependent on the desired and optimum logistics within the works, and the planned annual production. A potential expansion here should always be included in the calculations. Important points are the location of the plant and its links to transportation such as proximity to motorways, etc.

The Type of Paste Preparation Aggregates With respect to the different production methods, the types of the paste preparation aggregates can translate into completely different preparation machines and systems, depending on the end products and the raw material qualities to be used.

Layout of the Production Line (Plant Layout) The layout of the production line is dependent on automatic or semi-automatic production processes, which not only influence investment figures, but also requirements regarding capabilities and skills as well as the number of the operating personnel.

Selection of Additional Machinery Additional machinery serves to print, emboss or finish the materials. This depends on the end product's area of application and makes certain demands on space requirements and the number of personnel—depending on whether work is done inline or later with separately set up individual machines. Country-specific conditions such as concentration of personnel or full automation also play a role.

Specification of Equipping Machinery Equipping machinery is necessary for winding, quality control and packaging of the end products, which are prepared for sale. The scope of this work is heavily dependent on the end product to be manufactured because the buyer, for his part, specifies requirements and has certain

A. Giessmann, *Coating Substrates and Textiles,*
DOI 10.1007/978-3-642-29160-9_2, © Springer-Verlag Berlin Heidelberg 2012

delivery conditions. Today, controls are often waived at the time of receipt; the preferred policy being to fall back on manufacturer guarantees instead.

List Other Aggregates Also to be listed are other aggregates such as laboratory devices and auxiliary aids which are necessary for the operation of these aggregates and serve as aids for the operation personnel.

Potential Explosion Hazard when Solvents are Used A tendency exists to no longer use products containing solvents. However, because of their special characteristics, these products cannot yet be omitted completely. No fully satisfactory substitute has been developed to date.

An explosion hazard can exist in association with certain mixture ratios, as in the case of a PU coating, since an explosive gas-air mixture can be generated in the existing works or room temperature. An igniting spark could be generated by mechanical friction of rotating metals or by static charging of carrier papers and the coated substrates. For this reason, structural measures must be taken for explosion protection at the coating systems where work is done with solvents (BG Chemie 1989). This protection exists in the case of parts which rub together where aluminium is used, in the case of switches where low voltage is used, and is achieved in motors with the help of pressure encapsulation or flushing using compressed air.

2.1 Logistic Integration Within a Coating Facility

Both market studies and many other factors must be taken into account in the calculations in the case of new investments or when setting up a new coating facility. In addition to the technical requirements that a coating facility must satisfy, demands of a strategic and economical nature must also be observed. High productivity, reliability in meeting deadlines, minimum run times and quick reactions to changing customer needs and market situations can only be achieved by a consistent and seamless flow of information and materials between the system manufacturers, raw materials suppliers, substrate manufacturers, transporters and customers. These interactions can be described by the heading "logistic integration" . In the coating industry, logistic integration is understood as the networking and coordination of individual processes and procedures. Here, it is important to free oneself from the isolation and separation of individual function and task areas, and establish a system of so-called interdependence in solving/managing tasks (REFA 1992).

This is based on the following main considerations:

- Logistic integration connects sales and acquisition markets, as well as customers and coating companies.
- It consists of numerous links in a chain; i.e. processes which must mesh together in order to achieve optimum results.
- Strength and performance in this logistic chain is determined by your weakest link. The most important thing in logistic integration is not that individual links are particularly strong, but that none is weaker than the others.

Figure 2.1 illustrates the possible links in the chain of a coating company, whose controlled coordination and combination are essential for the success of a company.

The goal of logistic integration is to coordinate and optimise the flow of materials and information along all the links in the chain, and to remove sources of disruption in order to ensure the smoothest possible flow.

In Figs. 2.2 and 2.3 the functions of the individual links are illustrated.

The reason why smooth running is so important is because the customer makes his purchase decisions dependent on not only price and quality, but also on criteria such as terms of payment, courtesy conduct, deadline reliability or financing options. Thus, the following are essential attributes for performance requirements on logistic integration, (see also Hautz 1992):

- Technical performance/services
- Logistic performance/services
- Sales performance/services

2.1.1 Technical Performance/Services

This area includes all performance/services which deal with the technical possibilities found on the market. On one hand, the available machinery represents the "state of the art". On the other hand, the implementation of requirements pertaining to the products should be done in terms of engineering and know-how.

Functionality: Technical suitability, product characteristics (e.g. low fogging, re-action to repeated flexure, reaction to temperature, tear resistance, abrasion resistance) depending on the area of application.

Type: Function reliability in terms of technology, life span, maintenance in-tervals (e.g. precision of winding, exactness of cut edges, uniformity of thickness throughout the width and length of the material).

Technology: The newest developments in product and process technology, and flexibility (e.g. direct or indirect coating).

Innovation: Product ideas adapted to the market, product design (e.g. deep valley printing, two-tone effect, wash effect).

The increasing need for coated and laminated substrates can be explained by the fact that, thanks to these processing procedures, connected with the use of various chemicals, completely new product characteristics are created which are very inter-esting for the industry and the consumer. As it is, the automotive industry would be unthinkable without such products as airbags, side cladding made of plastic com-ponents, climate conditioned seats or covers. With things such as roof insulation, utilisation of solar energy, awnings, vinyl floor covering or wall cladding, the con-struction industry delivers new impulses. Coated and laminated substances are also increasingly in demand in the sport and household goods industries, where they are used in boots and tent materials, and in life vests and decorative outer claddings.

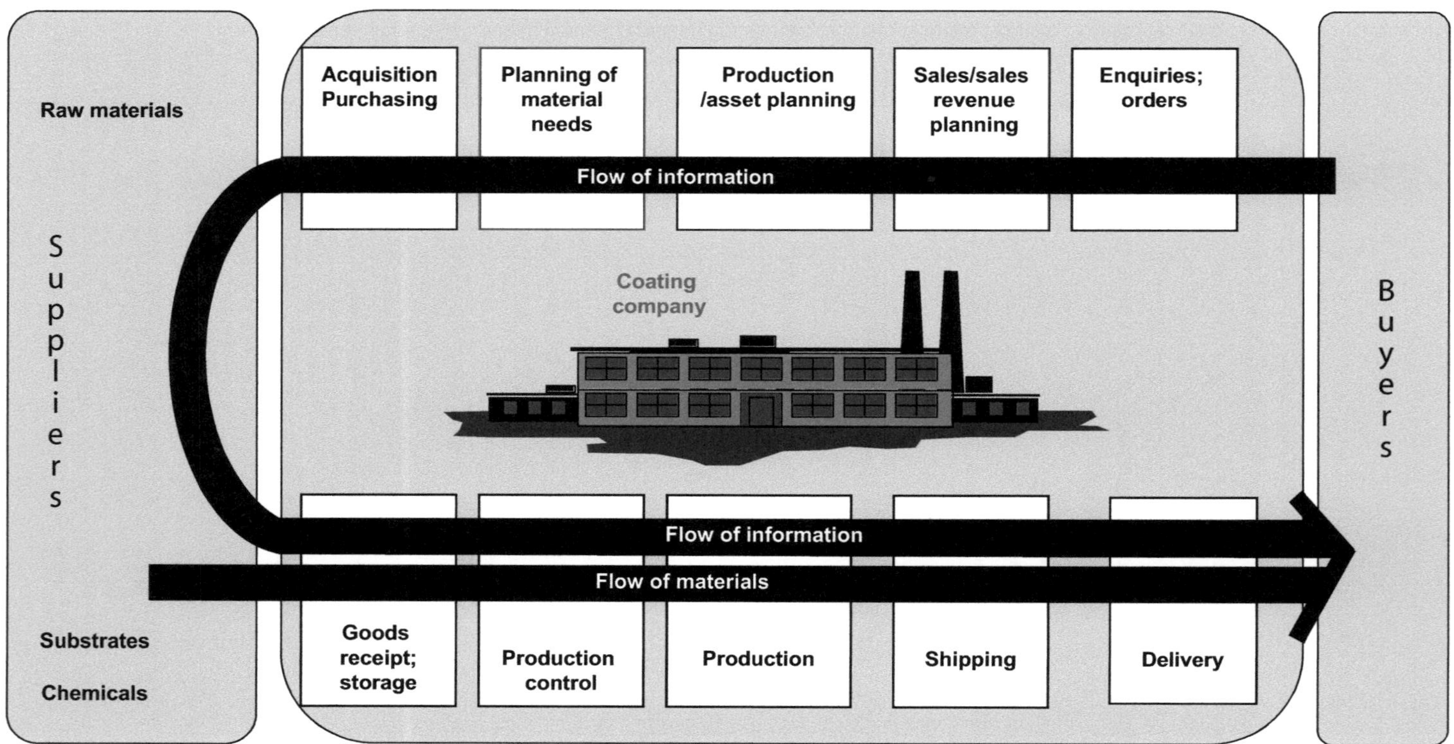

Fig. 2.1 Logistical chain. (cf. Zuendel and Partner 1992)

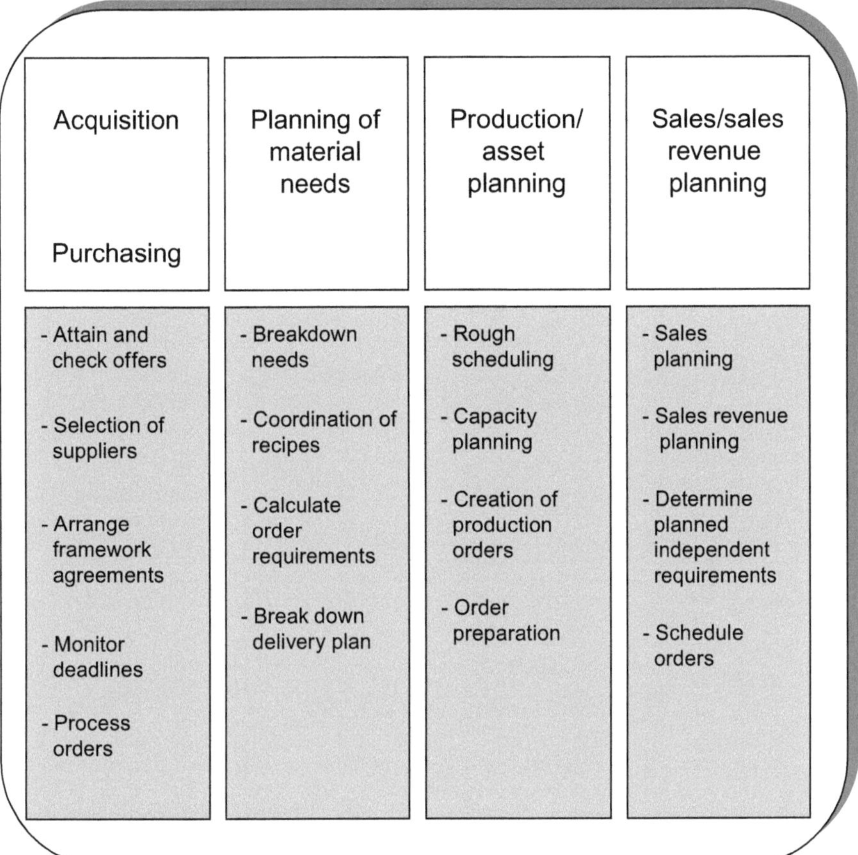

Fig. 2.2 Functional approach to logistic integration (1)

The business operation should maintain a suitable laboratory to help handle the increasingly shorter product life cycles and provide the innovative strength that a coating company must have. Laboratory facilities have the important task of taking responsibility for basic ideas and adaptations of formulae, without having to make use of the actual production equipment for this. This requires a great deal of knowledge of the various chemicals, whereby it should also be possible, by corresponding trials, to satisfy the demands made by the market in terms of characteristics and structure of the end product.

Only when one has the specific knowledge of the substrates to be used, and of the raw materials and the production methods, as well as the type of adhesion that is most suitable, can a product which is mature enough for the market be produced. Suitable laboratory machinery is necessary to examine each parameter precisely. On one hand, this includes the machinery for the manufacture of coating pastes and, on the other, the coating machinery itself. The latter is divided into either discrete or continuous

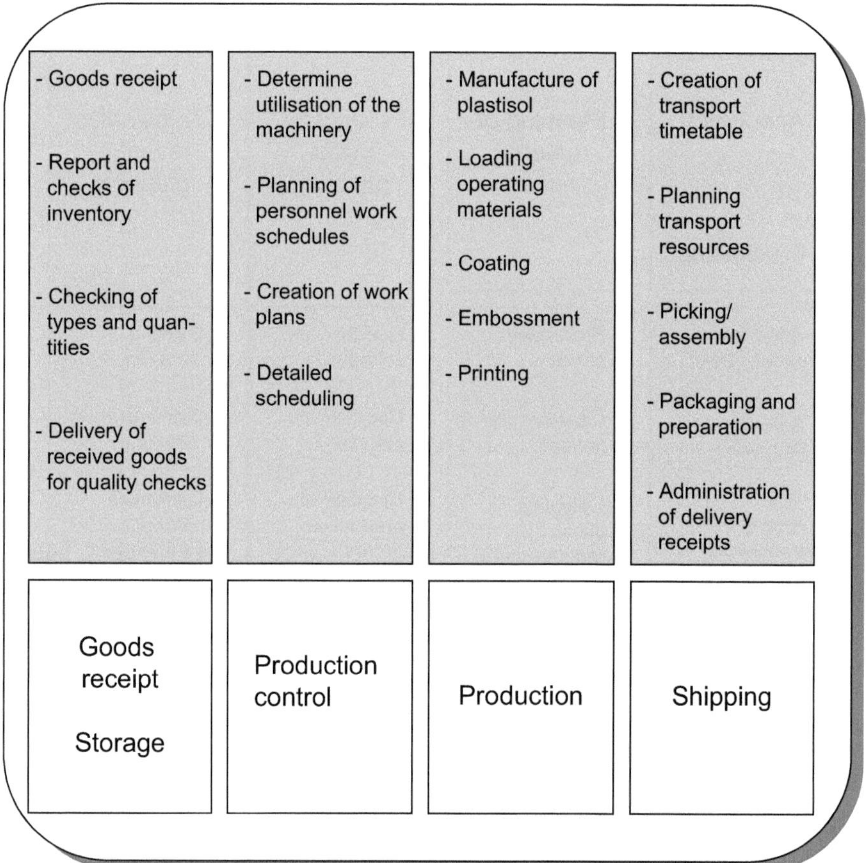

Fig. 2.3 Functional approach to logistic integration (2)

laboratory systems, which differ in terms of space requirements, working width, equipping of the application systems, as well as the required number of operating personnel. Table 2.1 provides an overview of the various laboratory systems.

The *discrete* equipment coats in the so-call drawer system. Here, the substrate to be coated is mounted in a pre-made frame and a wiper knife, or respectively, a rotating roller applies the coating. In this way, patterns can be produced in sizes up to a maximum of DIN A4. The devices themselves have the size of an oven. This system makes it possible to reach general conclusions concerning raw materials and reaction to temperature. Because of the fact that it is a discrete process, and because of the technical structure, the results can only be applied to the actual production process in a limited way.

In contrast, *continuous systems* are pilot systems which have a structure similar to the large scale industrial systems. Here, the work width varies between 400 and 1,000 mm. These systems are utilised to obtain realistic data for the production

Table 2.1 Characteristics of various laboratory systems for coating facilities

Characteristic	Discrete systems	Continuous systems	Special Deskcoater system
Space requirement	Approx. 1.5 m^2	Approx. 25 m^2	Approx. 5 m^2
Required personnel	1	3–5	1
Work width	Approx. 25 m^2	Approx. 40–100 cm	Approx. 50–60 cm
Ø Production speed	–	6–8 m/min	4–6 m/min
Reproducibility	Marginal	Yes	Yes
Flexibility	No	Marginal	Very high

process and reliable samples for customers. Disadvantages of these systems are the large amount of space required, the inflexibility with respect to various process methods, the very high cost, and the high personnel requirements.

Larger scale construction measures are necessary if different application systems are to be tested and adapted to the respective needs. It is even possible to place several aggregates in sequence. This naturally has the disadvantage that the distances to the tunnel change, and so they no longer reflect the layout of the production system.

The distance between the application of the coating and the entrance to the drying and gelling tunnel, in particular, is decisive for the quality of the coating. It is also necessary to pay attention to the type of application medium here: if a solvent-based media is used, then the distance should be as minimal as possible in order to prevent the solvent from evaporating and mixing openly with the atmosphere. Additionally installed slit exhausts or exhaust hoods are possible cautionary measures in this instance.

One new concept is laboratory systems which represent a symbiosis of both of the previously named versions. The Deskcoater—as pilot system—combines in one single system the compact construction of the discrete system with the possibility of applying the data gained to the continuous system.

The different application systems can be easily interchanged thanks to the modular add-on system. This way, the distance between coating and the entrance into the tunnel can be kept the same, and one receives reproducible data (Fig. 2.4).

Aided by the modular construction of the individual equipment features (coating module), each processing task can be individually readjusted. The application module versions range from the simple roller knife and multiple roller systems to the impregnation method and other production processes.

2.1.2   *Logistic Performance/Services*

The logistics chain connects all areas which are essential for a properly functioning coating facility and for a high degree of customer satisfaction. Logistics services must constantly be optimised so that organisation is not hindered by any individual link in the chain. Of particular importance here is the flow of information between the individual departments of the facility, as well as between the company and the customer.

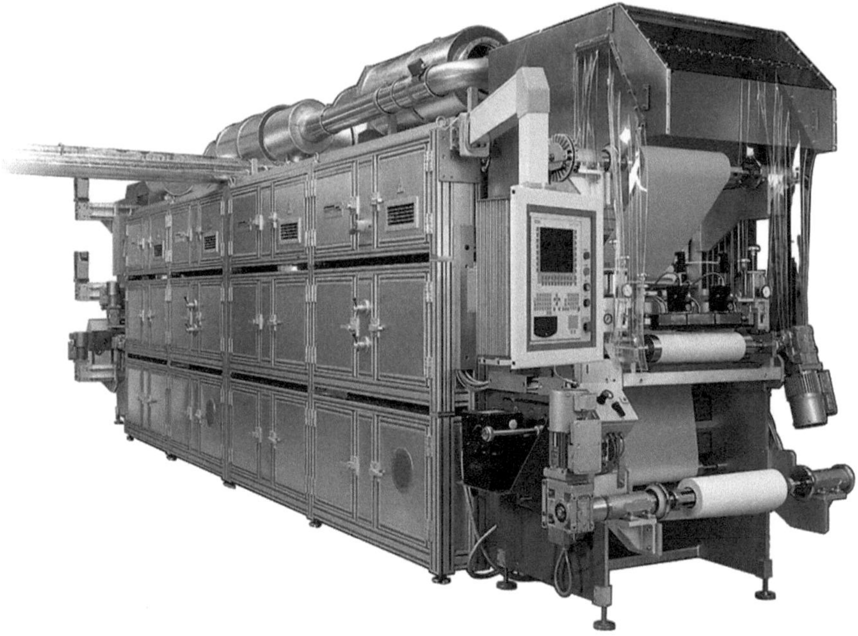

Fig. 2.4 Coatema Linecoater (factory photo)

Delivery time: Period of time between order confirmation and order comple-
                         tion.
Delivery reliability: Relationship between the stated and the actual delivery date.
Deliverability: Relationship between the delivery date desired by the customer
                         and the confirmed delivery date.
Delivery quality: Proportion of qualitatively and quantitatively correctly exe-
                         cuted orders.
Information: Quality of the information service or the flow of information
                         in all stages of the business transaction (Fig. 2.5).

2.1.3 Sales Performance/Services

All marketing activities that are aimed at introducing the right product to the market at
the right time are considered sales services; i.e. implementing the company strategy.

Spectrum: Breadth and depth of the services offered, as well as the
                              product palette.
Advice: Expertise, know-how, knowledge of the customer, care
                              service network.
Customer acquisition/care: Hotline, exhibitions, customer training, seminars.

Fig. 2.5 Front view of the
Coatema Deskcoater
(factory photo)

The level of priority of these factors is dependent on the individual needs and requirements of the customer, as well as the respective product. Here, it is evident
that in the case of so-called "grey products" (i.e. products with a similar technical
quality and price structure), out of all suppliers, preference is given to those whose
logistical integration is realised and executed the best. With the help of the principle of holistic design and coordination, lopsided partial services are avoided on one
hand, and on the other hand it is assured that the chain from supplier to customer is
stringently organised and structured, whereby the goals of the services are based on
the requirements of the market (Heeg 1991; Fig. 2.6).

2.2   Realisation of Logistical Integration

Products adapted to logistics allow acquisition, processing and transport processes
to be completed more quickly and synchronised more effectively. Optimisation of
the work processes facilitates decisive simplifications of organisational sequences
within the process chain. The use of specialists, or respectively, advisors, and the use
of their specialised knowledge increase the efficiency of any ventures undertaken.
Consistent information systems and a rapid flow of information are the conditions
required for the coordination of all process levels involved within a coating company.

Approaches for realising logistical integration are shown in Fig. 2.7.

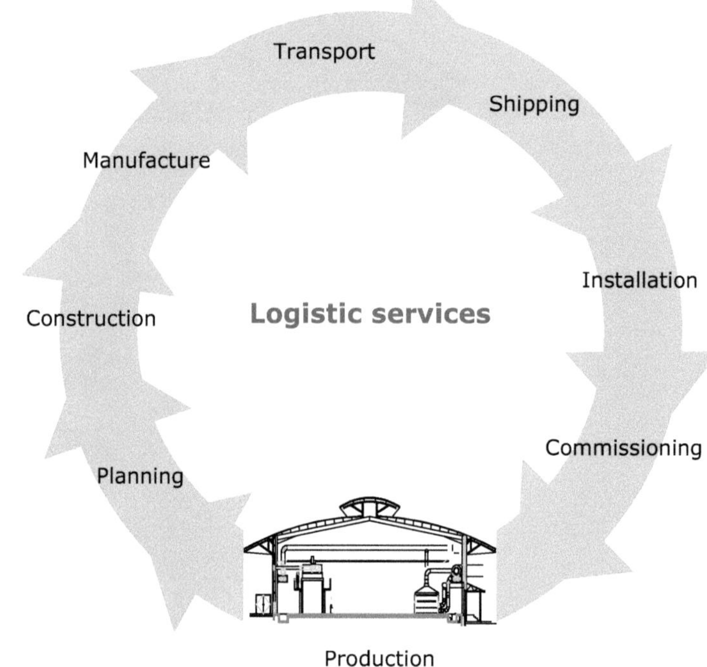

Fig. 2.6 Logistic services

If all links in the chain are to mesh, then the work in the departments must be mutually coordinated at the time of order receipt. That is why it is not only the responsibility of e.g. the purchasing department to acquire raw materials and thus assure the lowest possible cost of production. At the same time, the laboratory

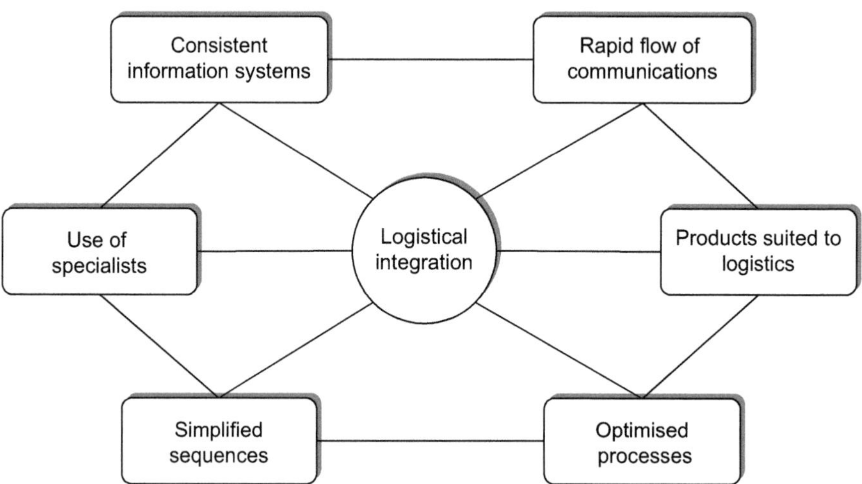

Fig. 2.7 Approaches to logistic integration. (Rammeisberg 1992)

must develop cost-saving recipes in order to save costs in mixing as well as in the ingredients. This is how the planning of the products and the production processes must play an important role in creating strong logistic chains.

The main task of modern logistic integration consists in making the transitions between the individual links in the chain more transparent in order to assure good coordination between the individual company departments.

Chapter 3
Basic Elements of Coating Systems

A coating system consists of many closely linked components which must be mutually coordinated. Figure 3.1 shows the structure and the working principle of a knife coating system.

The following basic processes can be carried out with such a system (Felger 1986):

- Unwinding
- Applying
- Drying
- Cooling
- Winding up

These processes include mechanical, thermal and chemical components. The roller, the essential component and basic element in a coating system, will be considered in detail first.

3.1 The Roller

The most important and most frequent component of a coating system is the roller. It comes in the widest range of models and performs several functions. As motor driven guide roller, it is responsible for direction changes in the web. As bearer drum (often non-driven) it provides the web with support, and as tension roller it ensures that the web is taut.

Another way in which the tension roller is used is as jockey roller. This is used, among other things, to coordinate the speed between two driven knife units. A swinging motion in vertical direction caused by the mass of the roller and additional mass pieces—today also pneumatic—triggers an actuator which adjusts the speeds of the knife units in relation to one another by accelerating or slowing the motors of the following units.

As hollow cylinder, rollers can also be used for process steps such as the heating or cooling of coating media and substrates which are yet to be coated. Here, special attention should be given to a large loop in order to exploit the heat exchange area available (Finke 1994). Guide rollers usually have a diameter of 80–160 mm at an

A. Giessmann, *Coating Substrates and Textiles,*
DOI 10.1007/978-3-642-29160-9_3, © Springer-Verlag Berlin Heidelberg 2012

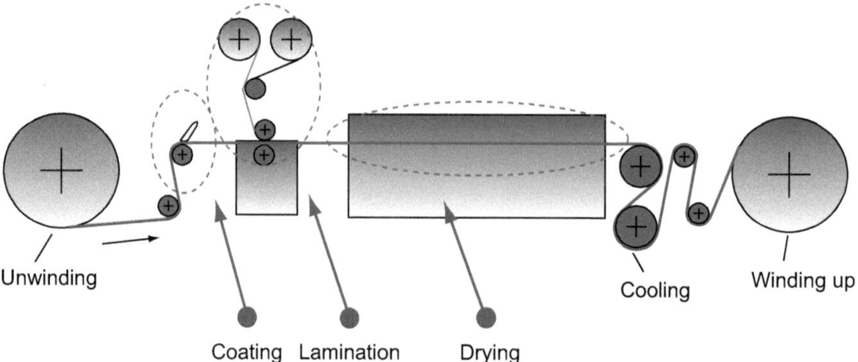

Fig. 3.1 Basic principle of a knife coating system

average working width of 1,400–2,000 mm. Heating and cooling rollers as well as knife rollers can have a diameter of 350–800 mm; those for winding up lie in the range of Ø 800–1,800 mm. Calender rollers, in contrast, have a roller diameter of 500–1,000 mm and a roller width of 1,250–2,500 mm.

Depending on the area of application, the rollers are made of steel (with chrome-plated surface), stainless steel, aluminium or are chill cast. At production speeds of > 60 m/min. the roller should also be dynamically balanced, especially when working with tension-sensitive materials.

The working speed of the coating system can be determined with the help of the roller diameter D and the roller speed nw:

$$V = \pi D n_w$$

With respect to the rotation speed, the roller—here as winding base for winding up or unwinding—has a special importance. Since the diameter increases constantly during the winding process, a continuous reduction of rotation speed—using a centre winder—is necessary to maintain a constant peripheral speed of the batch winder.

If a roller is used in a roller pair, deformation, squeezing and transport operations take place. Influencing factors for the roller gap and the operations which occur there include the roller area, the roller diameter, the amount of pressure applied, bowing of the rollers, the choice of materials, as well as the same or opposing direction of rotation.

The deflection which occurs in every roller disrupts proper functioning in many cases. For this reason, the construction can be altered to limit this drawback (Ungricht 1992):

1. Convex shape of outer surface of the roller
   The roller is ground to a convex shape; i.e. the roller diameter is larger in the middle than on the ends.

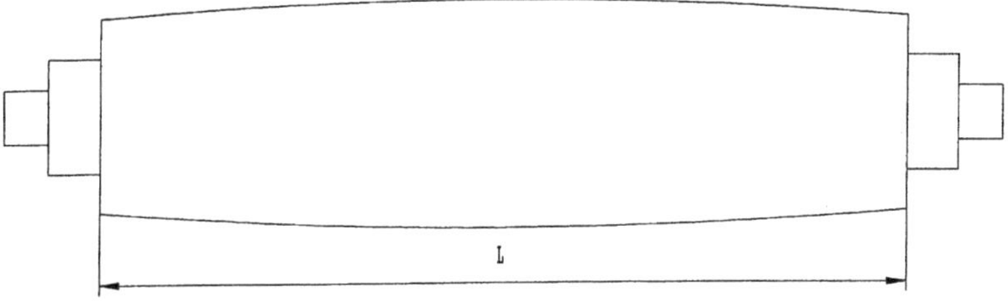

2. Reduction of unsupported spans
   The interior of the roller is reinforced. One way to do this is via a hydraulic liquid.
   In this case one would refer to this as "swimming roller".

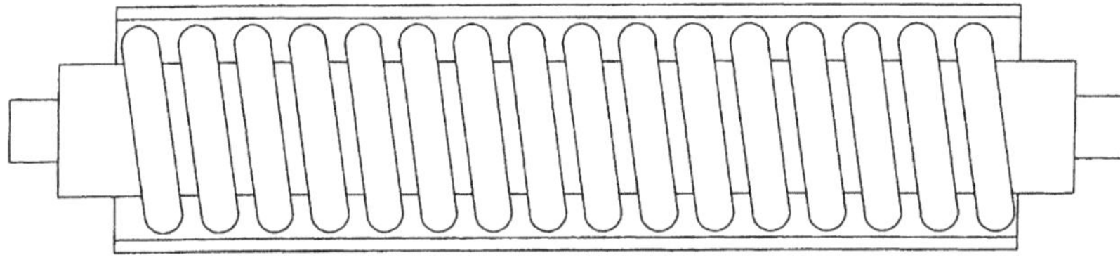

3. Enlargement of the roller diameter

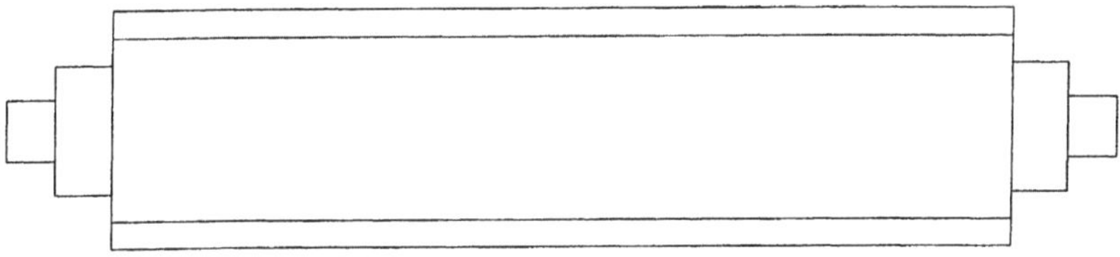

4. Setting at an angle or crossing of the pair of rollers to maintain a uniform gap
   between the rollers (cross acting).

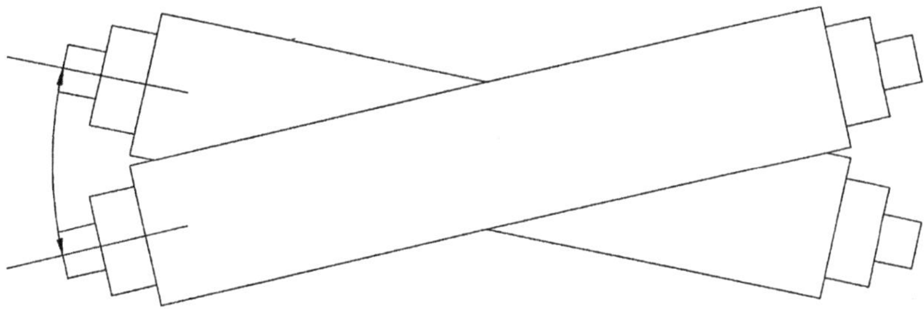

5. Application of counter-pressure to the roller necks of the pair of rollers
   The rollers are pre-tensioned and receive a negative deflection (roll bending) as a
   result.

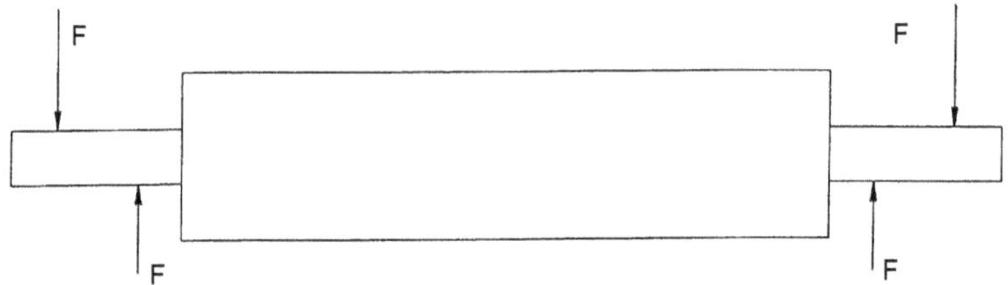

6. Exterior reinforcement of the roller by several small support rollers

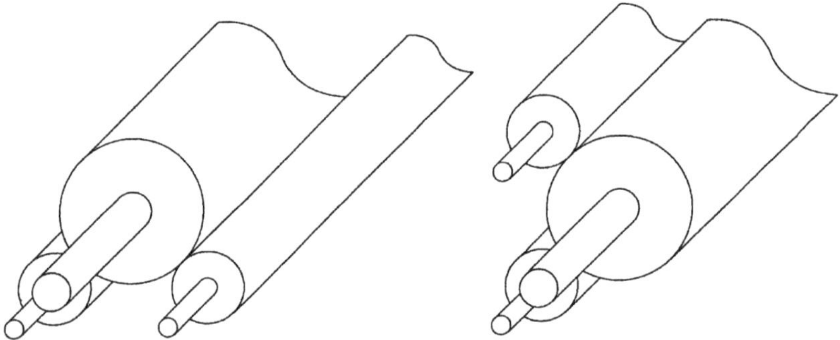

Knife rollers are subjected to particularly high demands with respect to surface
quality and consistency of diameter.

In addition to the rollers described, there are other rollers which, because of their
surface, apply specific effects to the products. One differentiates between so-called
engraved rollers, embossing rollers and print rollers (Dornbusch 1994).

Engraved rollers are rollers which are used for final lacquer coats and for minimal
application thicknesses such as e.g. adhesives or lacquers. Here, the most common
rollers are finely engraved with pin points. A so-called mesh number indicates the
number of points which, in a regular arrangement on the corners of equilateral
triangles, are situated in a straight row of the closest nearby points at 1 in 2.54 cm.
The number of points per cm^2 can be calculated as follows:

$$\text{Number of points/cm}^2 = \frac{\text{Mesh number}^2}{2.54^2} \times \frac{2}{3}$$

Embossing rollers are used to change the structure of the surface; i.e. with their help
the material (e.g. imitation leather) is given its corresponding grain. In contrast, in
the case of print rollers, the colours and motifs are applied to the product. Examples
of this are table cloths, curtains or flooring patterns.

Table 3.1 Roller diameter for the coating of different product groups

Product	Tension (10 N/m)	Roller diameter (mm)
Imitation leather	80–100	100–120
Floor covering	130–150	160–200
Tarpaulins	120–150	140–180
Clothing	80–100	100–120

Table 3.2 Overview of various shear rate ranges

Action	Shear rate range D (s^{-1})	Examples
Sedimentation of fine particles in a suspension	$10^{-6}\ldots10^{-4}$	Paints and lacquers
Sedimentation of coarse particles in a suspension	$10^{-4}\ldots10^{-1}$	Ceramic suspensions
Running caused by surface tension	$10^{-2}\ldots10^{-1}$	Paints, printing inks
Dripping subject to gravity	$10^{-2}\ldots10^{-1}$	Spreadable paints, coatings
Pipe flow	$10^{0}\ldots10^{2}$	Pumped liquid
Mixing, stirring	$10^{1}\ldots10^{3}$	Process engineering
Spraying, squirting, wiping	$10^{3}\ldots10^{6}$	Coatings
Rolling	$10^{4}\ldots10^{6}$	Printing inks, coatings
High speed coating	$10^{5}\ldots10^{6}$	Spreading machines for paper

The material for the roller surface is essentially dependent on two factors (Wannagat 1995). The climate conditions of the location of use (rust risk) on the one hand, and, on the other, the chemicals to be used (solvent) are decisive. The roller surface material consists of the following, depending on how the rollers are used: chrome-plated steels (basic guide rollers), aluminium rollers (heating rollers, cooling rollers), rubber-coated steels (doctor knife rollers, squeegee rollers, drawing rollers) or respectively chill cast rollers (high precision application rollers and calender rollers).

Different products require different tensions. It follows that the respective roller diameters must be sized accordingly. Sizes based on experience—with respect to four common product groups—are listed in Table 3.1.

Since the focus in the textile industry is on knife coating, the different coating components employed by this method will be described in more detail.

3.2 Application Equipment

Viscosity is a basic factor in the behaviour of substances to be processed, such as suspensions, dispersions or pastes in general. It is useful as gauge of resistance to the flow of a liquid in laminar flow (Table 3.2). The formula is

$$\eta = \frac{\tau}{D}$$

τ = shear stress
D = shear rate

Table 3.3 Overview of various viscosity ranges

Substance (temperature) at $T = 20\,°C$	Dynamic viscosity η (mPas)
Air	0.018
Acetone	0.32
Petroleum ether	0.54–0.65
Water 20 °C	1.00
0 °C	1.79
40 °C	0.65
Ethanol, alcohol	1.2
Milk, cream	5–10
Motor oil	100–500
Castor oil	Approx. 1,000
Honey	Approx. 10,000
Syrup	Approx. 1,000–10,000
Molten plastic	10^4–10^8
Silicone rubber	10^5–10^8
Molten glass	10^{15}

Shear stress is defined as the force K with respect to the area F, whereby the direction of force is parallel to the working surface. K is used to move two layers against one another. The shear rate indicates the speed at which the layers can be pushed against one another.

This way both the production speed as well as the coating gap are included in the choice of application system is to be considered. With application by a doctor knife, with a coating gap of 2 mm and a production speed of 0.5 m/s, the usual shear rate $D = 250$ increases tenfold to 2,500 s^{-1} (Table 3.3).

When calculating the viscosity and the behaviour of the paste at various shear rates, so-called viscosity curves result:

Newtonian Viscosity remains constant independent of the shear rate, e.g. water.

Shear Thickening Viscosity increases with rising shear rate, e.g. starch.

Shear Thinning Viscosity drops with rising shear rate; with dropping shear rate it forms in a predicable way; e.g. dispersion paint.

Thixotropic Viscosity drops with rising shear rate; with dropping shear rate it slowly forms again; e.g. gel, lacquer, yogurt.

Depending on the application medium and viscosity of the coating paste, numerous appliances have been developed and proven their worth for coating with plastisols, dispersions, or suspensions. Differentiation is made between knife and roller application systems.

3.2.1   Knife Systems

The description "coating with a doctor knife" is a specialised term used to describe the process of coating when the knife works against an underlay, a roller, a table

Fig. 3.2 Roller knife

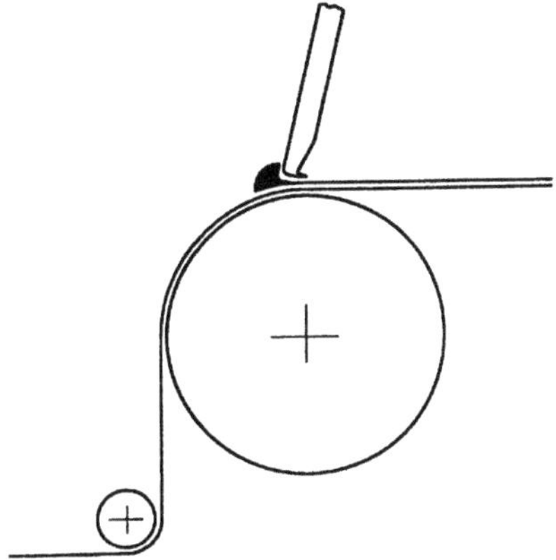

or the substrate itself. Application by knife can thus be understood as a spreading device which is fixed in place and whose total width is as wide as the web. The wiping mechanism has the shape of a knife, which is usually made of certain types of steel or, less frequently, hard rubber.

In the context of conventional methods of coating, differentiation is made between the following systems of application using these wiping knives (coating with a doctor knife):

- Roller knife
- Air knife
- Rubber belt system
- Support knife
- Table doctor knife
- Spiral knife
- Case knife

3.2.1.1 Roller Knife

In the case of roller knives, the knife works against a roller. This can be a rubberised roller, a chrome-plated steel roller, or, in the case of precision applications, a chill cast roller. Precision applications are understood as application thicknesses of 5–8 g/m^2 which require a concentricity of the roller of 0.002–0.004 mm (Fig. 3.2).

The so-called *rubber roller*—a rubberised steel roller—is commonly used when uneven substrates are due to be coated. The rubber roller is increasingly replacing the rubber belt, which was frequently used in the past, and is only used today for coating stripes (awnings) (Lamalle 1991). The minimum application accuracy here lies at approx. 50–100 g/m^2. That is why its use only makes sense for rough and unshorn fabrics.

The *chrome-plated steel roller* is the most economical roller; its application precision lies at approx. 25–50 g/m^2.

Fig. 3.3 Air knife

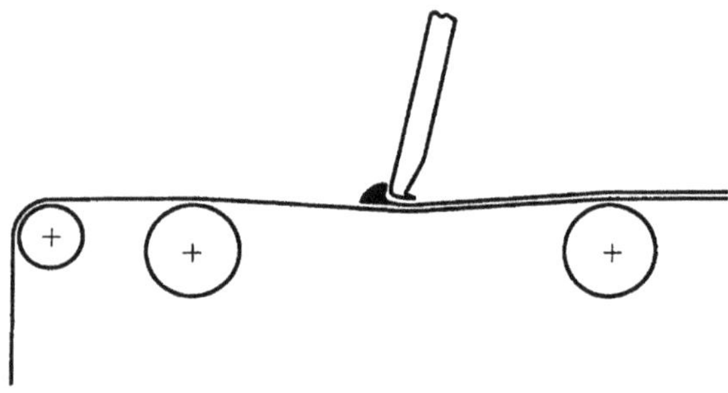

When executing transfer coating of textiles and interlaced fabrics—e.g. to manufacture imitation leather—*precision rollers* are used with which applications of 5–10 g/m^2 can be achieved, whereby the upper limit is virtually unrestricted.

3.2.1.2 Air Knife

The air knife is formed from two rollers with a gap of approx. 25–40 cm. The web to be coated is fed over two rollers (coming from below because of the necessary loop), and the knife applied either centrally or near the roller with the web looped around it. The counter-pole is the air. The paste is now placed in front of the knife, and the application is determined only by the tension and the speed of the web. The air knife system is extremely useful in the case of the thinnest of coatings, as it is used in the case of rain materials, surf and parachute materials, or other polyamide weaves such as nylon and Perlon.

The transferability of settings is one problem associated with the air knife. These settings are usually parameters gained from experience, gathered from trials. The amount to be applied is adjusted by the tension, and especially by the speed.

All the spreading methods mentioned function with the help of a doctor bar to which the spreading knives are fastened. The doctor bar must be mounted completely free of vibration and must never bow (Fig. 3.3).

The connection-free construction of the doctor bar is also of extreme importance. The gap forces can be countered with the help of additional counter-pressure mechanisms, in order to prevent lifting (dancing), which would lead to so-called chatter marks.

In order to prevent long production stops such as in the case of changes to coating operations or the replacement of a spreading knife, knife bars are equipped with two or also three (special constructions) different knives.

Knife Coating Machines Today, knife coating machines most commonly combine both the roller and air knife systems by making the doctor bar infinitely adjustable by a linear guide over or behind the coating counter-roller.

Forms of Doctor Knives The doctor blades for air and support knives have a small and rounded chamfer of approx. 1–2 mm width. They serve to prime light fabrics

Fig. 3.4 Knife coating machine (factory photo Coatema)

and non-woven materials with plastic dispersions. In the case of the knife edge of the roller, rubber belt and/or table doctor knives, the chamfer width lies between 3–15 or 20 mm, whereby these taper off towards the rear surface. Here, one speaks of a so-called back-taper which prevents the paste from collecting behind the blade, so no rupture of the film is caused. In general, the chamfer thickness behaves analogous to the quantity applied; i.e. the larger the quantity applied, the larger the chamfer thickness to be chosen.

The thickness of the coat in the spreading process is dependent on

- The distance between the doctor knife and the substrate
- The type of substrate to be coated
- The type and form of the doctor knife
- The viscosity of the paste
- The speed of the web (Figs. 3.4 and 3.5)

Yet another way to make adjustments is by changing the angle between the web and the doctor knife, which also has an effect on the shear force which affects the paste.

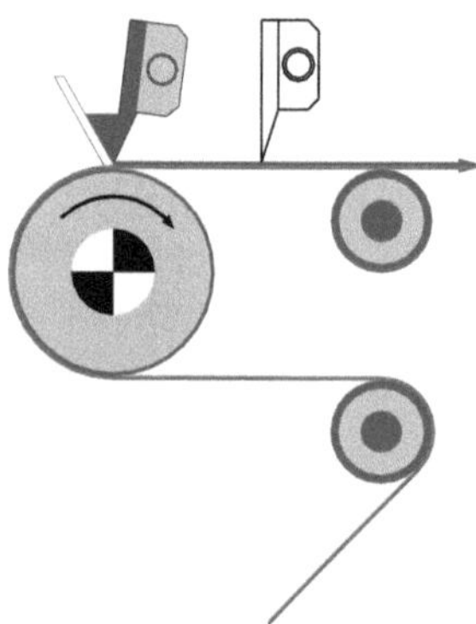

Fig. 3.5 Schematic diagram of the knife coating machine

3.2.1.3 Rubber Belt System

In the case of the rubber belt system, the doctor knife works against a rubber belt set taut between two rollers, whereby one roller is motor driven and the second roller serves as tension roller. Arranged midway between these rollers is another small roller which works as counter roller for the doctor knife. This ensures that a certain amount of adjustment to the gap can be made, despite the elastic behaviour of the rubber belt. The rubber belt rotates with the help of the tension roller and the additionally attached support roller bears the load that is created by the pressure of the doctor knife. This prevents the rubber belt from sagging (Fig. 3.6).

Today, the rubber belt doctor knife is used to manufacture striped awnings. Here, the doctor knife is combined with a box which is divided up into sections which can hold differently coloured pastes. This way, multi-coloured stripes can be made without an additional print operation. The important thing here is that the separating walls of the individual sections themselves function as lengthwise doctor knives, in order to ensure a clean separation of the individual colour strips (Figs. 3.7 and 3.8).

3.2.1.4 Support Knife

The support knife consists of two knives positioned opposite one another. It is used to spread the paste on two sides in cases of two-sided coating or of open/sparse fabrics, and paste that is pressed through the fabric.

3.2.1.5 Table Doctor Knife

The table doctor knife is becoming more and more obsolete because it has a rubbing effect on the underside, which can lead to excessive strain on substrates. This means that non-woven fabrics, which are usually more sensitive to tension than woven material or paper, can tear even when only a small amount is applied. The table doctor knife has now largely been replaced by the roller knife as a result.

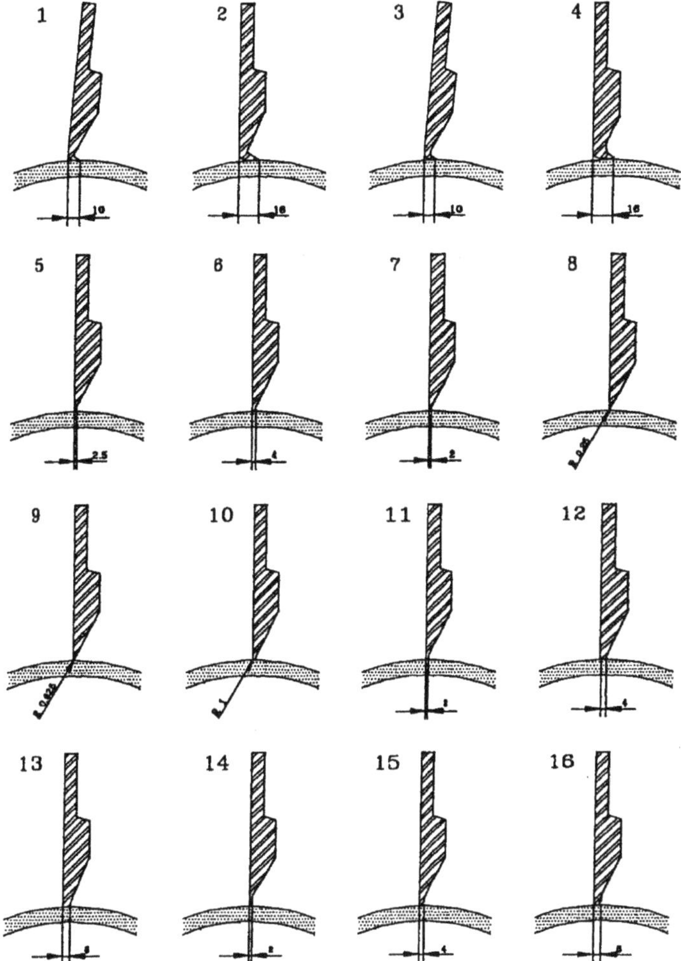

Fig. 3.6 Illustration of doctor knifes with various chamfers

Fig. 3.7 Rubber belt system

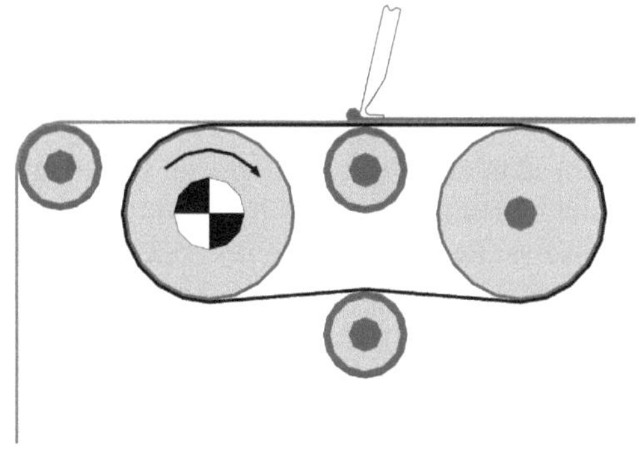

Fig. 3.8 Support knife

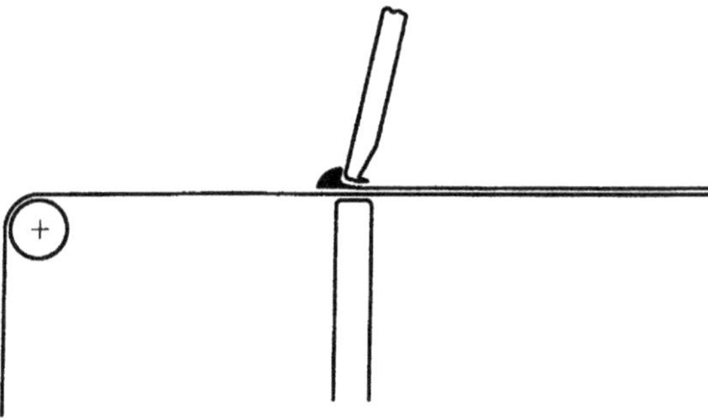

Fig. 3.9 Table doctor knife

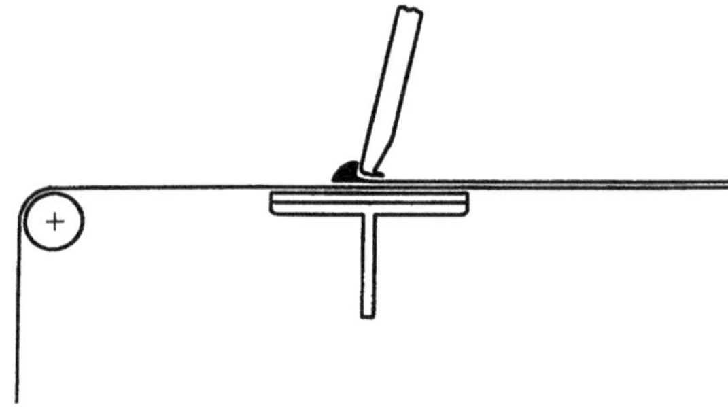

Fig. 3.10 Universal Revolver
Coater

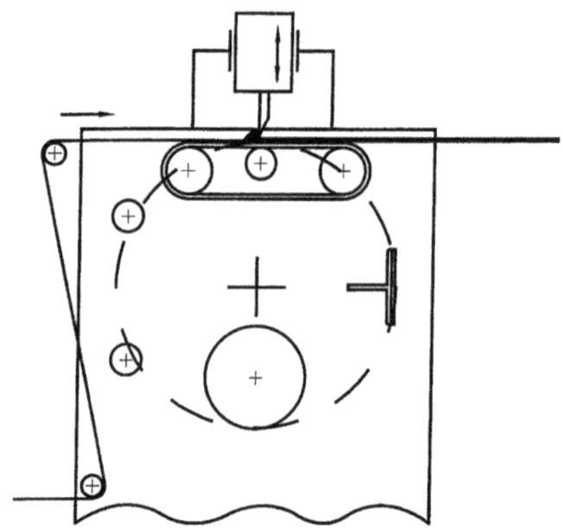

In addition to the various doctor knives, there is also a unit that combines the four knife systems presented, in one knife coating machine. This machine—called the Universal Revolver Coater—is conceived in such a way that the roller knife, rubber belt knife, air knife and table doctor knife are arranged in a rotating frame around the centre. The small construction size, the flexibility and the quick changeover are a few of the advantages of this construction (Figs. 3.9 and 3.10).

Fig. 3.11 Diagram of a spiral knife

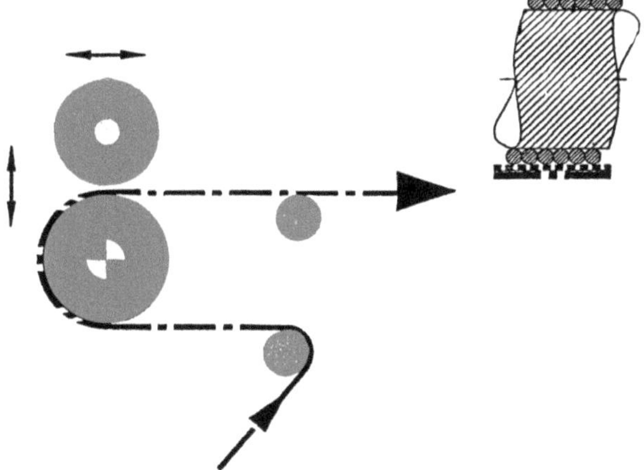

3.2.1.6 Spiral Knife

The spiral knife—also known as the Mayer rod system—is a core wound with a wire that is positioned fixed or also rotating in relation to the web. Here, the diameter of this wire determines the amount applied, because the doctor knife lies mechanically on the material. There are also other systems available that minimise other mechanical stress on the substrate. A typical application thickness is 1–75 g/m^2. Today, the spiral knife is occasionally used in visible areas of foil coating. This system can also still be found in the paper industry, to finish paper with an adhesive, in order to create self-adhesive products.

3.2.1.7 Case Knife

With the help of the case knife, thixotropic media that is filled into a closed box can be applied, running counter to the direction of the substrate.

Because of the fact that the substrate is already subject to contact with the medium because of the size and the application opening of the box, from the standpoint of time the possibility of joining together is longer, so there is a higher possibility of adhesion between the substrate and the medium. The front knife applies a kind of pre-dosing which, because of the counter direction of the roller (reverse drive) and because of the roller gap itself, results in the actual fine adjustment (dosing of the amount to be applied). Here, depending on the chemicals, work can be done with either smooth rollers or engraved rollers (Figs. 3.11 and 3.12).

3.2.1.8 Comma Bar

The comma bar system is a combined application system that unites the characteristics of a standing knife system with those of the roller application system. A

Fig. 3.12 Case knife

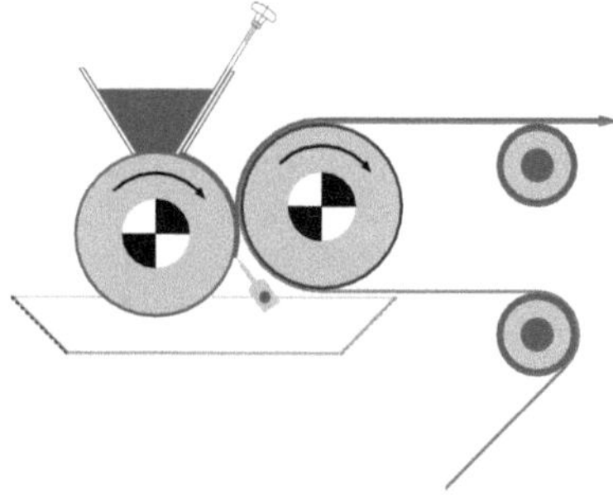

Fig. 3.13 Comma bar

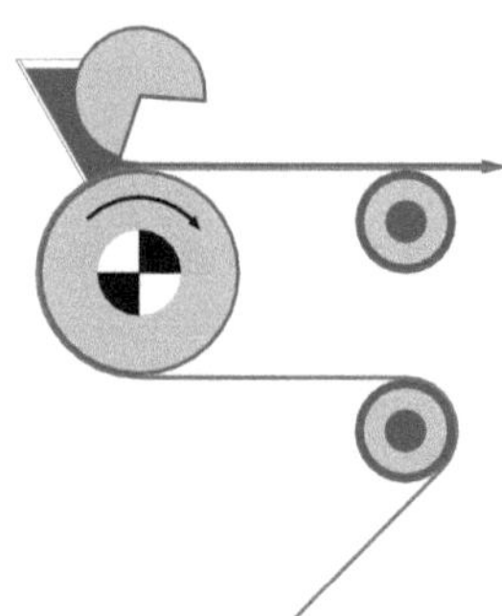

relief-ground roller is used instead of a doctor knife. The comma bar system is used particularly for coatings which demand deep penetration (e.g. non-woven fabrics). Because of its special geometric form, the comma bar has also proven itself as ideal system for chemicals with a very short working life, and thus tend to dry quickly on the application system (Fig. 3.13).

3.2.2 *Roller Application Systems*

Roller application systems do not apply the coating using doctor knifes, but rather with rollers alone. The only use that the attached knives have is to wipe off the excess paste and to smooth down the applied medium. Differentiation is made between four different roller application systems:

- Reverse roll coater
- Contra coater
- Roll coater
- Kiss coater

Fig. 3.14 Reverse roll coater

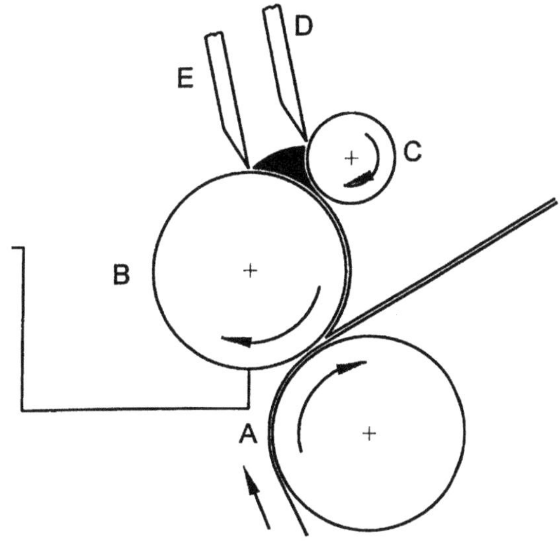

3.2.2.1 Reverse Roll Coater

The system is called the reverse roll coater because the application roller is arranged to run counter to the sheet of material. The reverse roll coater is a roller application mechanism with rollers arranged in an S-shape (Fig. 3.14). The coating (B) and the dosing rollers (C) are upstream above the rubber impression roller (A). A delimitation knife (D) and a wiping knife (E) form the gap and ensure clean application.

In this system each roller has its own separate drive, which is usually a DC motor. The speed of each roller relative to one another can be varied by approx. 50 %. The direction of rotation can be reversed. Because of the many adjustment options the coating is achieved either through the size of the gap, the peripheral speed or the relative speed. Additionally, a wiping effect can be achieved if desired.

The reverse roll coater is suitable for working with well-flowing pastes and other thin coating media. Coatings of 2–3 g/m^2 are possible here. The main area of use of the system is the lacquering of coated fabrics such as apron cloth, camping articles and table cloths, as well as to apply adhesives to self-adhesive films. Because of the high shear forces that occur, shear-thickening pastes, especially those with a high viscosity, cannot be processed.

3.2.2.2 Contra Coater

The described system is also available as four-roller reverse roll coater. It is called a contra coater, but it works with an addition supply roller (F), which is partially submerged in a reservoir and then passes the paste attained there to a coating roller (Fig. 3.15).

Fig. 3.15 Contra coater

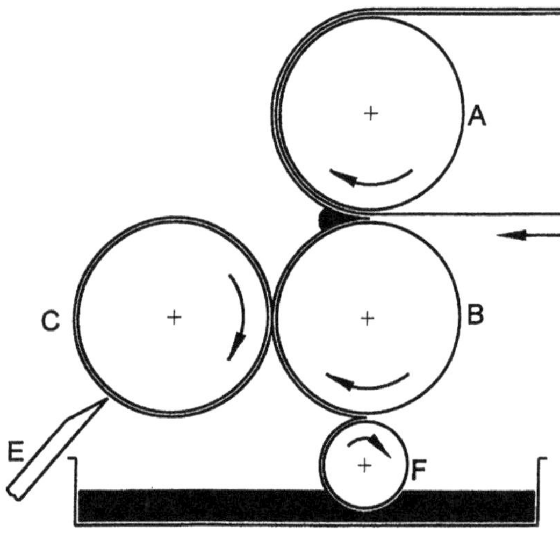

Fig. 3.16 Roll coater

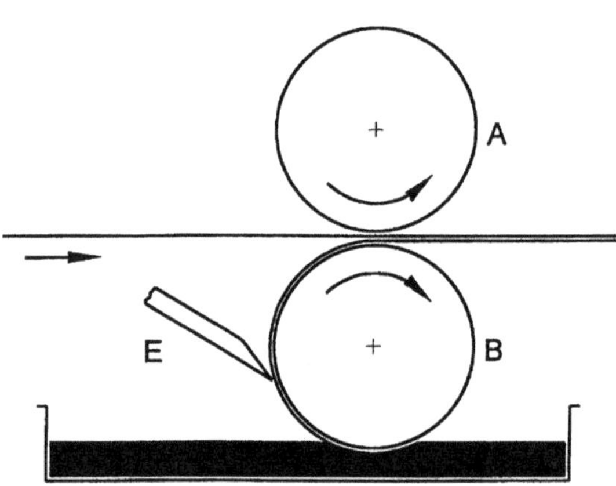

3.2.2.3   Roll Coater and Engraved Roller

In the case of the direct roll coater, which also goes by the name of kiss coater, the carrier material is transported between two rollers. The coating roller (B) transfers the paste from a reservoir to the substrate, which is pressed against the application roller by a second counter-roller, the rubber impression roller (A) (Fig. 3.16). This method of coating was developed to apply extremely thin films (10–$30\,g/m^2$), e.g. for packing papers and wallpapers.

The engraved roller works according to the same principle; here, instead of the smooth application roller, an engraved roller is used. The doctor knife (E) is set very close to the roller to wipe away excess coating material from the surface (Fig. 3.17). All that is left over is the amount of coating determined by the engraved surface.

Fig. 3.17 Engraved roller for direct coating

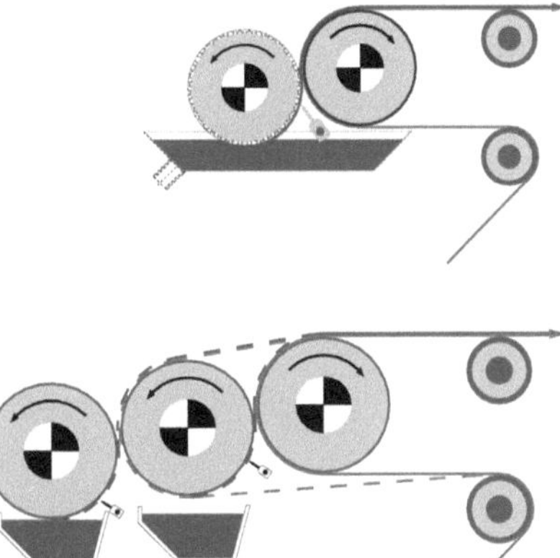

Fig. 3.18 Engraved roller for indirect coating

Depending on the surface of the product, a decision is made whether coating is to be applied to the substrate directly by the engraved roller or indirectly with the help of a rubber roller.

Direct application with an engraved roller is chosen in the case of smooth surfaces. However, should the surface structure be uneven or have higher thickness tolerances, damage to the substrate is avoided through a transfer of the bath using the rubber roller. This ensures that the engraved image is fully transferred. In the case of a system which is indirect because of the layout of the system, modern systems also allow direct coating so that two systems can exist in one (Fig. 3.18).

3.2.2.4 Kiss Coater

The kiss coater works according to the same principle as the direct roll coater, only that instead of the counter-motion roller, two guide rollers and a specific tension are responsible for application. This technique is suitable for high coating speeds for priming coats, and requires pastes with low viscosity.

A classic kiss coater has a chromed roller as application roller, which obtains the coating material from a reservoir and transfers it to the substrate. It should be possible to adjust the direction of rotation and the roller speed independently of the production speed, in order to achieve the best possible flexibility.

3.2.2.5 Rotation Screen Printing

This type of coating system is used to coat partial areas. It is used in situations where materials in a web form need to be printed. If specific structures must be applied to a substrate, they can be integrated in the screen as design elements. An internal doctor knife distributes the paste and presses it through the template. The angle the knife is

Fig. 3.19 Kiss coater

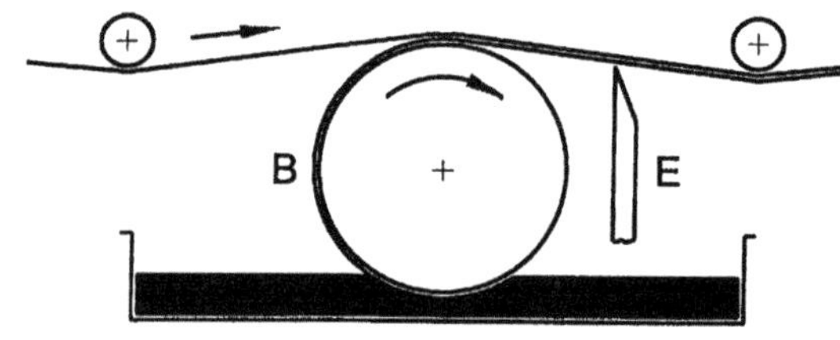

Fig. 3.20 Rotation screen printing

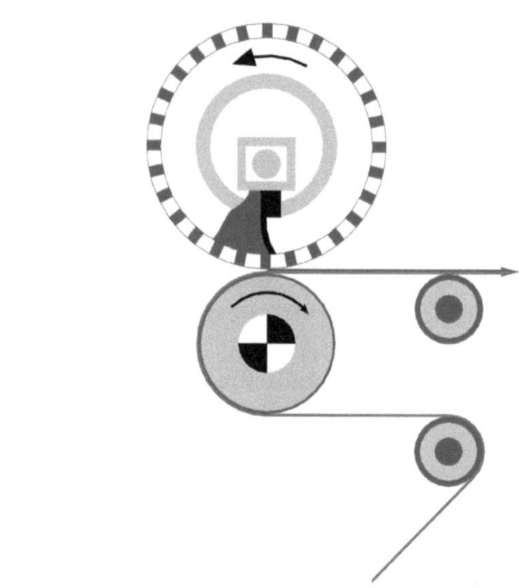

set at and its position are decisive for influencing the quality of the print. The screen itself can be destroyed very quickly if the setting is incorrect.

In terms of reasons for use, the rotation screen printing system can be compared to some extent with an engraved roller system. Here, too, in the case of very thin coatings, an appropriate screen can be used, so that because of their adhesive strength, the applied points can spread out into a cohesive film after contact with the web. Viscosity is an important factor for achieving a good application result. Here, the viscosity range is very small.

Attempts have also been made to use the screen as full-area application system, which has, however, succeeded in only a few cases. This is how the so-called "Whisper blade" came into being, which was supposed to finally wipe the applied coating of a partial area into a cohesive film. But because the "Whisper blade" only has a simple construction, it could not achieve the quality of a conventional knife system. At the end of a coating process, not only the "Whisper blade" but also the screen must be cleansed. That is why the knife system is preferable in cases of full surface application (Figs. 3.19 and 3.20).

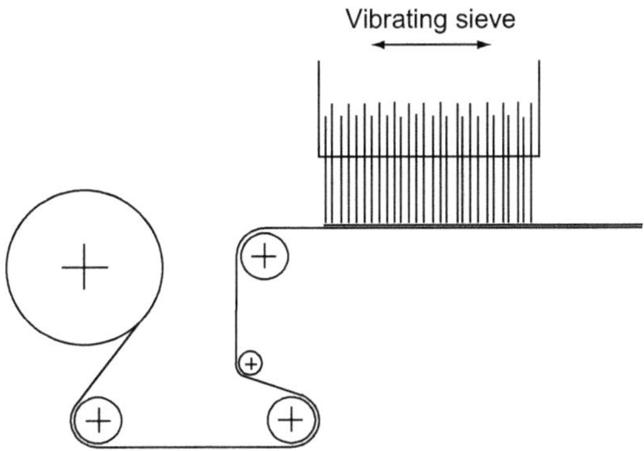

Fig. 3.21 Powder coating using a vibrating sieve

3.2.3 *Powder Coating*

During powder coating the thermoplastic coating material is sprinkled uniformly on the medium using volumetric dosing, and subsequently melted, cooled and smoothed. This ecologically harmless method does not require any solvents.

Application accuracy in this system lies at $\pm 2\,\%$ at a width of up to 5 m. The coat thickness with respect to good reproducibility lies at 10–$1{,}000\,g/m^2$.

This method is used for:

- Clothing textiles
- Fixable fabrics that are suitable for stiffening
- Non-woven linings
- Transport and conveyor belts
- Laminating in general

Various thermoplastic materials such as LDPE, HDPE, PP, PA, PES, EVA and PU are used.

Today, in addition to the vibrating sieve, needle rollers are often used. The roller receives the coating material through a funnel and uses the rotation movement to sprinkle the specified amount of material over the gap onto the textile. A brush that moves on its axis ensures that the amount to be applied is properly dosed (Figs. 3.21 and 3.22).

3.2.4 *Hot-Melt Method*

The hot-melt or melt roller method is derived from the principle of calendering. In contrast to the hot-melt system, calendering consists of only two to a maximum of three rollers. The thermoplastic is applied to the two heated rollers and melted there. A film is formed in the gap between the rollers which is applied to the substrate by a pressing roller, thus coating the substrate. After that it is cooled and rolled up.

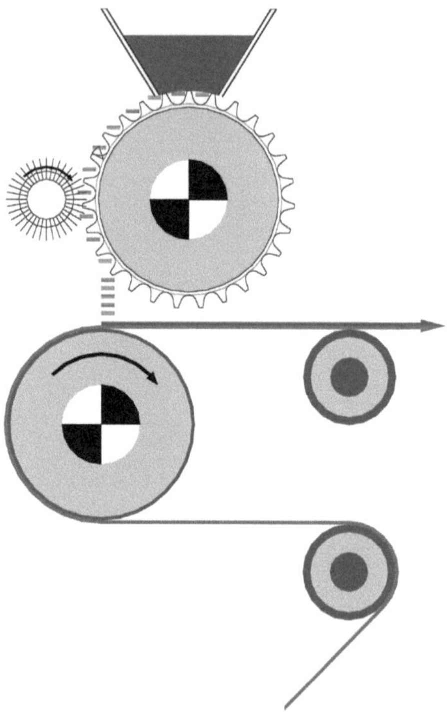

Fig. 3.22 Powder coating using a needle roller

A standard hot roller application system is based on the three-roller engraved application system which is used in the printing industry for gravure printing. It consists of three rollers: a transfer roller, an application roller, and a dip roller. The application roller can be replaced with an engraved roller if an extremely exact coating weight is required. Compared with slot die or powder coating application, the structure of the surface of the substrate to be coated is not an important factor in a hot-melt roller application system. Open as well as closed substrates can be coated. This covers a wide range of different coating weights.

In the gap between the dip roller and the transfer roller, the hot-melt material is conveyed by the rotation of the rollers to the transfer roller as film. The weight of the coating is defined by the setting of the gap between the dip roller and the transfer roller, as well as by the gap between the transfer roller and the application roller and the substrate. The speed also influences the weight of the coating.

A second substrate can be laminated to it by integrating a laminating station. When using thermoplastic hot-melt systems, the distance between laminating and the application of the raw material should be kept as to a minimum because the cooling of the adhesive has an influence on the viscosity, and thus on the penetration of the adhesive.

The hot-melt method is particularly relevant during the application of highly viscous adhesives with subsequent laminating. In addition to the roller application system, the hot-melt method also includes application via the slot die system or using

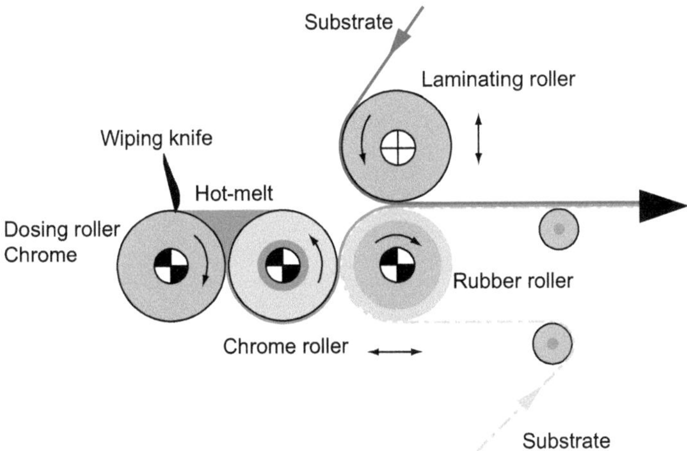

Fig. 3.23 Hot-melt with roller application system

the already discussed methods such as spray application, rotation screen printing or powder coating (Fig. 3.23).

In the slot die system, the material to be coated is melted in a hot-melt tank system and conveyed by a level monitor to the dosing unit along a heated hose. The textile carrier material is coated via a slot die nozzle by a pump system.

The thicknesses for hot-melt usually lie at 8–950 g/m^2, for slot die nozzles up to approx. 60 g/m^2.

Careful handling of the system is essential for hot-melt coatings. The coating head must be positioned very close to the substrate to prevent changes to the temperature caused by the movement of cold air. A drop in temperature can lead to fluctuations in viscosity, irregularities, the formation of strands or a hardening of the system, resulting in a poor coating effect.

The nozzle can be flexibly positioned above the transport roller. Manual or motorized adjustment allows the nozzle to be moved horizontally or vertically, as well as enabling a change of angle. The main advantage of the nozzle is the enclosed feeding of adhesive. This protects the hot-melt adhesive against contact with oxygen and moisture in the air, from the point where it is melted right to the point where it is applied. This way, even reactive hot-melts have no opportunity to crack or react due to contact with moisture.

The coating weight is determined by the pump power, the working speed, the distance between the nozzle and the substrate, as well as the tension of the material. This process requires the substrate to be highly stable, otherwise shear strain could occur. The enclosed system is the reason why high speeds are not a problem when coating with hot-melt adhesives (Fig. 3.24).

The main areas of use of hot-melt coatings are for adhesive application and/or in paper technology. Additional advantages of this method are the relatively high coating densities which result from roller printing, and thus high resistance to wear.

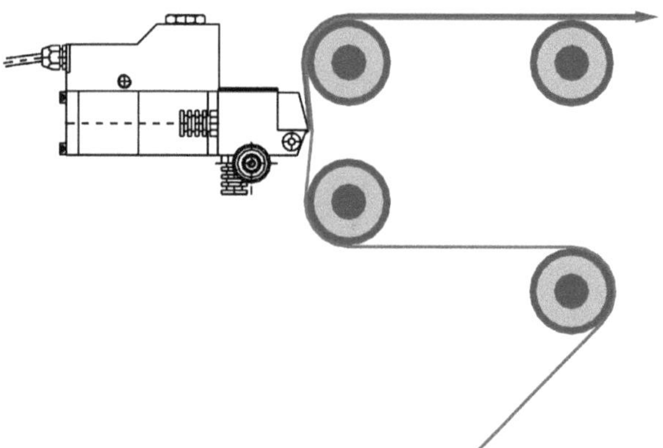

Fig. 3.24 Hot-melt nozzle

Such systems can process: M-PVC, S-PVC, PVF, PU, LD-PE; HD-PE, PB, PP, ABS, as well as EVAC-VC, CPE, CSM, PA and rubber EPDM, EPM and CR. Gluing is defined as the joining of parts by adhesion and cohesion whereby the substrates undergo no structural changes. Meltable adhesives are polymer based, thermoplastic glues that liquefy at temperatures between 80 and 220 °C and harden again when they cool. They consist of 100 % solids and they are applied in a liquid state without the use of water or solvents. Precise dosing and adjustments are possible when the melting, the pump system and the application technology (slot die nozzle, hot rollers or powder coating) are all taken into consideration. Depending on the process, in comparison to the processing of dispersion and solvent-based systems, hot-melts have shorter joining and fixing times (Tables 3.4 and 3.5).

The layout of the system and the process are determined by the hot joining technique, the binding and reaction time (working life), the inner adhesion, the production speed and the coating weight, but also by the hot-melt material itself.

The following characteristics must be taken into account with hot-melt adhesives:

Viscosity	Flow behaviour of the melted mass, depending on the temperature
Heat stability	Time during which the melted mass can be held at a certain temperature without compromising the dosage units
Open time	Maximum possible period of time between application and the actual lamination
Initial adhesion force	Force which a hot glue seam can withstand following a specific time (within the open time)
Setting time	Calculated from the moment the adhesive is applied, this is the time it takes to achieve maximum seam strength
Temperature stability	Temperature range in which the glued area can be subjected to a certain ongoing load

Also included in the processing of hot-melt systems is the preparation and conveying of the melted mass. In order to be able to convey melted glue it is necessary to melt

Table 3.4 Advantages and disadvantages of hot-melt coatings

Advantages	Disadvantages
• Environmentally friendly because waterless and solvent-free adhesives are used • Low coating weight • No drying necessary/minimal energy consumption • Less thermal stress on substrates • High production speed • Permanent or non-permanent adhesion possible	• Switching the type of adhesive requires intensive cleansing of the entire system (tank, hoses, nozzle, etc.) • Limited use, depending on the softening point of the thermoplastic • Limited working life in reactive systems • Brief reaction time (time between coating and laminating procedures) • Price

Table 3.5 Characteristics of different types of hot-melt

EVA (ethylene vinyl acetate)	PA (polyamide)	PUR (polyurethane), e.g. moisture induced crosslinking type
• Large softening and temperature range • Heat resistant up to 100 °C • Sensitive to steam and water • Economical	• Softening point higher than EVA • Temperature resistance up to 150 °C • Sensitive to steam and water • Quite expensive	• No influence by temperature after crosslinking (hot or cold) • Destruction of the adhesive above 150 °C • Boil-proof • Very expensive

it by heated metal surfaces or rods in an anti-adhesion coated aluminium/cast tank, and to pump it through electrically heated hoses to the actual application machinery.

The pumps are adjusted to match the coat thickness. Experience has shown the following values to be useful:

Gear pumps	[5–150 bar]	(10 g–250 kg)/h
Piston pumps	[5–250 bar]	(10 g–270 kg)/h
Extrusion pumps	[5–150 bar]	(5 kg–50 kg)/h

3.2.5 *Injecting*

The method of plastic injecting is used primarily in the automotive industry for underside protection and sealing seams, or in the field of textile carriers for the application of lacquer or adhesives. A spray gun is used to coat. Compressed air is used here to force the liquid plastic through the nozzle of the spray head. It is not only possible to work with solutions or dispersions, but also with previously melted plastic, in order to be able to spray the so-called melted mass. In the textile industry, this type of application is usually used in the case of technical textiles such as insulation mats or PU foams. A further area of use is the application of glues in stations that perform pure lamination.

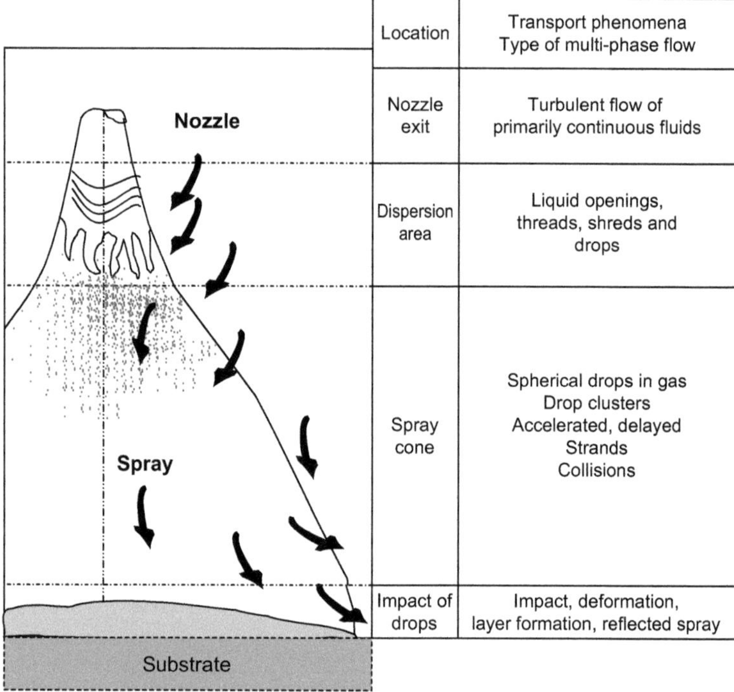

Location	Transport phenomena Type of multi-phase flow
Nozzle exit	Turbulent flow of primarily continuous fluids
Dispersion area	Liquid openings, threads, shreds and drops
Spray cone	Spherical drops in gas Drop clusters Accelerated, delayed Strands Collisions
Impact of drops	Impact, deformation, layer formation, reflected spray

Fig. 3.25 Diagram of flow areas during spray application

In spray coating, four different flow areas can be differentiated:

1. Liquid flow in the nozzle, or nozzle exit
2. Zone of the spray or film disintegration directly after the nozzle
3. Spray cone
4. The area in direct proximity to the substrate

Special demands are made on the rheology of the pastes in order to achieve uniform application and good atomisation.

The pastes must lend themselves well to spraying or extruding. The decline in speed that exists in the nozzles can be approximated by the following formula:

$$D = \frac{32 \times V}{xt \times d3}$$

Declines in speed of up to 10^5 s^{-1} are achieved with today's airless sprayers.

At a minimum pastes should be shear thinning or must generally have a yield point in order to be suitable for spray coating. Of course, this type of coating system is being used less and less frequently, because the losses—up to 30–40 %—are very high. Additionally, the problem of the spray cone should not be underestimated (Fig. 3.25).

Fig. 3.26 Spray application

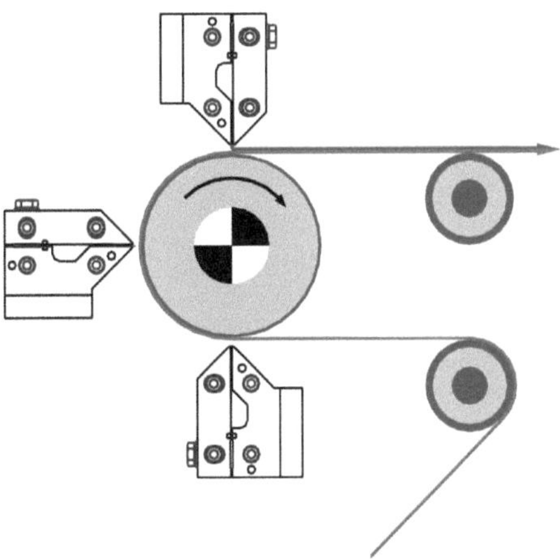

3.2.6 Slot Die and Pouring Technologies

Slot dies or pourers are enclosed coating systems which consist principally of two metal sheets fixed against one another. By inserting masks between these two plates, the width of the coating can be determined by their opening width, and the coating weight by their thickness. The masks are cut precisely from polymers or metal (usually stainless steel) materials by laser cutters.

The main purpose of slot die technology is the application of a very uniform coating film, precisely defined in terms of width. S die and pouring technologies are used predominantly for low to medium viscosity coating substances.

The design of the slot die can differ widely. The principle elements defined are the feed groove, the length of the exit slot, and the lip of the slot die. The coating substance is pumped into the slot die from a reservoir through a pump system and hoses and distributed uniformly in the feed groove over the full width before being homogenised by the exit slot and coated through the nozzle lip.

The weight of the coating is defined by the mask in the nozzle as well as the pump and substrate speed.

The determination or selection of the pump technology must consider the fact that the coating substance must be conveyed to the nozzle in a uniform way. Micro-gear pumps, diaphragm pumps, and pressure tanks are preferred here. When conveying the coating substance to the nozzle, special care must be taken to ensure that no air is introduced to the nozzle. In an optimally controlled coating system, the pump speed is controlled by the software in such a way that when the system speed changes, and thus also the substrate speed, the pump speed is automatically adapted (Figs. 3.26 and 3.27).

In addition to the design of the slot die, also the position of the slot die, the adjustment angle, and the coating characteristics have an influence. Here, differentiation is made in positions between the 6 o'clock and the 12 o'clock settings. The viscosity and the required application weight play a role here.

Fig. 3.27 Principle
construction of a slot die

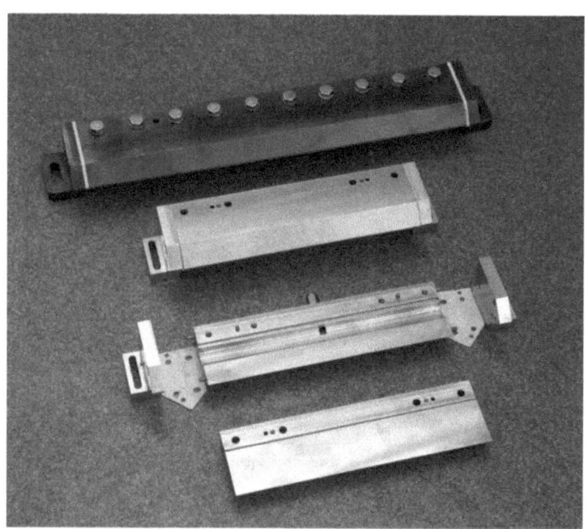

Fig. 3.28 Position of the slot
die

Depending on the distance between the slot die and the substrate, differentiation is made between coating with contact, and so-called curtain-coating (coating without contact). When the substrate runs slowly, work is usually performed with contact to the substrate, and when the substrate runs faster, without contact. Here, the term "pouring" is often used for the coating process; this means that the coating substance is poured onto the substrate or a transfer carrier like a curtain (Fig. 3.28).

The advantage of slot die or pouring technology is the uniform application of a closed film, regardless of the substrate to be coated.

A counter roller with very high concentricity is also crucial to the achievement of high accuracy. As an enclosed coating system, slot die technology has the following advantages:

1. Optimum use in an environment resembling a clean room, because the coating substance has no contact with the surrounding atmosphere and so the conditions under which it is applied are very pure.

2. The vaporisation of solvents is prevented because the coating substance, coming from a sealed container, is never exposed to the atmosphere during transport through the hose system and nozzle.

The areas where slot die systems are used are diverse, and are increasing constantly along with the need for high quality coatings. Among others, slot die systems are used for:

- Very thin, uniform films
- Application in the shape/form of a stripe
- Very uniform coated films, regardless of the type of surface of the substrate
- Solvent-based coating substances
- Situations demanding a high degree of cleanliness

Even a simultaneous double-sided coating is possible. Slot die systems can be used horizontally as well as vertically. The broad range of applications makes slot die systems an interesting coating system whose use is increasing.

Cleaning involves either pumping a cleanser through the opening, or the system is dismantled for more rigorous cleaning. This can be done very easily by unscrewing the two slot die plates, and requires minimum effort.

3.2.7 *Dipping and Impregnating*

Dipping is understood as the coating of three-dimensional objects. This means, a fully manufactured object such as e.g. a glove is dipped in a coating substance and, after removal, dried or finished in another way. A similar principle is applied also to sheet goods in cases where the coating is not only intended for the upper and lower surfaces, but also the side edges are to be coated. Depending on the viscosity and the material used, the material to be coated is also fully penetrated.

In the case of impregnating, the web is drawn through a dipping reservoir—also called a foulard (French)—which contains the coating substance. After the web has emerged from the dipping reservoir the excess coating substance is either downright squeezed off by a pair of rollers (e.g. in the case of dying textiles) or, as stationary system, or while rotating counter to the web (e.g. coating of glass fibre) the pair of rollers squeezes off the excess coating substance, down to the desired coat thickness. A chrome and a rubber roller are used for squeezing which can be adjusted in pressure towards one another. In the case of stationary rollers or rollers running counter to the web, the application amount is determined mainly by the gap between the two rollers. As illustrated in Fig. 3.29, appropriately designed doctor knives can replace stationary rollers. The arrangement of the squeezing roller pair is shown horizontally in Fig. 3.29; however, vertical arrangements can also be found in practice.

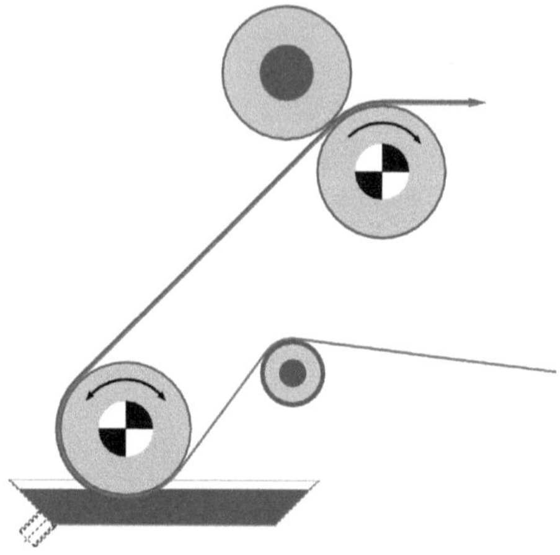

Fig. 3.29 Dipping method using squeezing roller or doctor knives

3.2.8 *Point and Double-Point Coating*

Point coating techniques were developed from screen printing methods. In the case of point coating, the coating substance (i.e. the adhesive) is pressed through the template onto the substrate. In this case, the amount applied is determined by the template used.

3.2.9 *Foam Coating*

Foam coating is to be understood as the generation of polymer films with a pore structure on sheet-like substrates. Here, the foam can have either open or closed pores, and depending on which polymer is used and the nature of the recipe, it can be adjusted for any kind of material properties, from soft to hard. Foams are characterised by the following properties:

- Grip effects
- Weight and material savings
- Heat and cold insulation
- Noise insulation
- Padding effects
- Absorption of impact forces (packaging) (Fig. 3.30)

Three manufacturing methods exist for foams:

Foaming by Chemical Foaming Agents Chemical foaming agents are products which decompose at higher temperatures and thereby release gases. This process is irreversible and usually occurs exothermally in a certain temperature range.

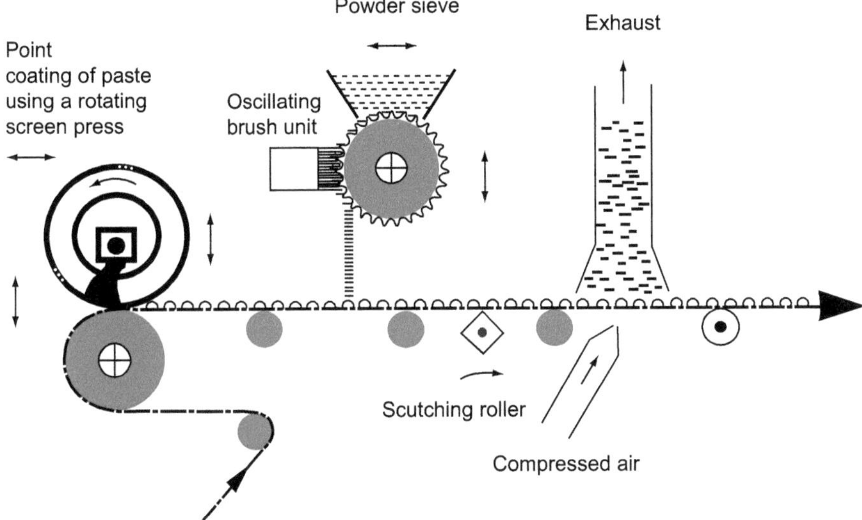

Fig. 3.30 Illustration of the double-point method

Foaming by Thermal Foaming Agents Here, thermally expandable microcapsules are mixed into the substance to be coated. The microcapsules have a particle size of 10–20 µm and have an expansion rate of up to 6 times that of their original condition.

Foaming by Mechanically Introduced Air Compressed air is incorporated in dispersions with the help of foam generators or foam mixers.

3.2.10 Other Methods

Two other coating systems and their methods are addressed as follows. In addition to spread coating, calendering is of particular importance.

3.2.10.1 Calendering

In the case of doctor knife or roller application, liquid coating substances are used which subsequently harden in a drying process (gelling). In the case of calendering the exact opposite occurs: the coating substance is heated.

The calender method is used mainly to manufacture foils, table coverings, shower curtains and imitation leather out of emulsion PVC.

A calender consists of multiple rollers which are arranged in a certain order relative to one another in a common frame. The rollers are usually arranged vertically above one another.

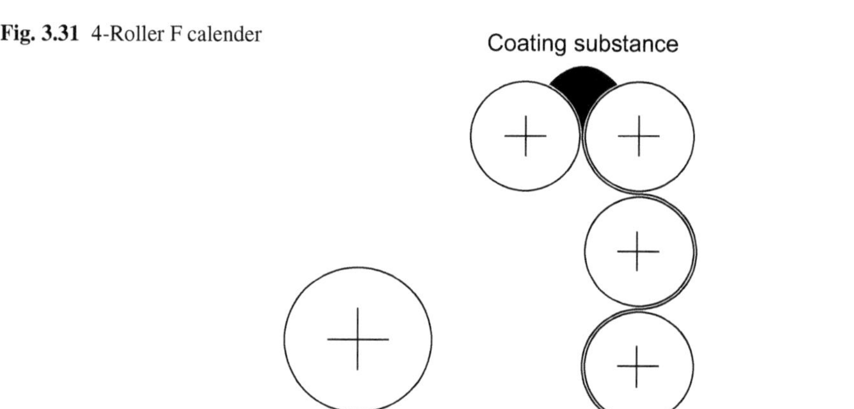

Fig. 3.31 4-Roller F calender

The prepared compound (i.e. the plastic which has been plasticised under heated conditions) is forced through steel rollers which have an increasingly narrow gap, at a temperature of approx. 200 °C, until the desired thickness has been achieved. The calender is continuously fed by extruders. The extruder is a screw or shaft that rotates in a heated cylinder and has one or more coil-like links.

The process steps in calendering comprise the following points:

- Preparation of the PVC medium in the mixer with all the ingredients such as plasticisers, pigments, additives
- Homogenisation of the plasticisation in the units situated upstream from the calender, such as kneaders, extruders, or roller mechanisms
- The forming process on the actual calender (Fig. 3.31)

In the textile coating process using a calender, the textile carrier material is pressed onto the hot film by the final calender roller or by an additionally mounted laminating roller. It adheres without the necessity of an additional coat of adhesive.

3.2.10.2 Coagulating

Coagulation is understood as the hardening of a semi-liquid mass in a precipitation bath with appropriate chemicals. Here, from the standpoint of the machinery, the production principle of the dipping method is adopted. This means that a plastic dissolved in a solvent—usually DMF—coagulates in a solid form as soon as it is thinned using a non-solvent. The coagulation method makes it possible to achieve a micro-porous PU coating for the manufacture of imitation leather.

After the substrate has been unrolled from an unwinding station, it is impregnated by a polyurethane solution in a reservoir. Excess coating mass is squeezed off in a squeezing mechanism consisting of a pair of rollers. Instead of an impregnation process, it is also possible to apply the coating in a dipping tank that is filled with a mixture comprised of solvent and nonsolvent. Here coagulation is triggered because the solvent is forced out of the solution by the nonsolvent. The dissolved plastic remains stuck in and on the carrier material, and this way binds itself to the micro-porous layer.

The repeated wash and squeezing procedures in the downstream units remove the remaining solvent. The DMF is removed by exploiting the concentration gradient to the wash bath concentration, of the solvent mixture in the micro-pores of the imitation leather. The number of required wash operations is mainly dependent on the following points:

- Production speed
- The concentration of the solvent
- The temperature of the wash bath

Today, coagulation systems are operated in connection with solvent recovery systems.

3.3   Material Transport

Material transport refers to how the materials move through the system. It must be exactly matched to the respective substrate and the layout of the system.

3.3.1   Winding

In the coating industry, so-called batch winders are responsible for unwind-ing/winding up. Usually large batch winders are used with diameters of 1,800 mm and more, which allow longer piece lengths than small batch winders (Ø 600 mm). The high meterages considerably shorten setup times.

Large batch winders are classified as either fixed or enclosed systems. Work is done with safety chucks in the case of open systems. The version of the winding rod can vary, and the winding rod be taken out, and this way be transported. Rollers and cores have established themselves as winding elements. Rollers are smooth, cylindrical bodies. In the case of cores, however, cardboard tubes of various diameters can be placed on cores and anchored by mechanical or pneumatic means, for example.

In the case of enclosed roller systems it is always only possible to transport the entire winding unit. The open system has established itself widely because it is easier and more flexible to use.

Another property that differs in the systems is the method of winding. There are surface and centre winders. In the case of surface winders the rotation speed is constant ($n =$ const). The degree of hardness at which the material is wound is determined by the difference in speed of the two rollers. The centre winder uses a different method. Here the rotation speed varies ($n =$ var.) and the diameter changes.

Care must be taken with all winding units that the material is fed without any folds. The side where the material is rolled up must always have a higher tension

Fig. 3.32 Factory photo, Coatema large batch winder

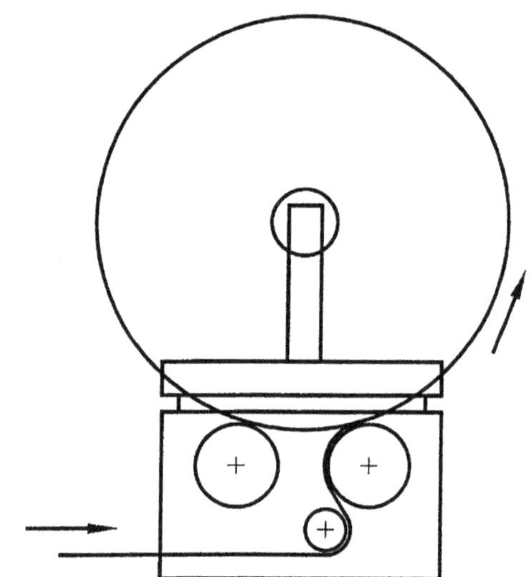

Fig. 3.33 Ascending batch winder

than the side where the material is being unrolled. These units are usually driven by hydraulics or a motor with additional winding calculators (Figs. 3.32 and 3.33).

In addition to the abovementioned winders, there are also the ascending batch winders and turret winders. Ascending batch winders are surface winders whose winding rollers can be given different kinds of coverings, depending on the type of material, in order to achieve the best possible take-up of the substrate.

Turret winders can be used to wind smaller meterages such as in the manufacture of flooring. On a turret winder, two small batch winders are mounted on one axis so that the second, empty batch winder is pivoted into the place of the first, after the first is full.

In this respect, the following formulae are useful to calculate the batch winder diameter D with a known, wound web length of L, or respectively, to calculate this

Fig. 3.34 Factory photo, Coatema turret winders

winding length with a known batch winder diameter; here, S stands for the thickness of the material:

$$D = \left(4 + \frac{SL}{\pi}\right) + d^2,$$

$$L = n \times \frac{(D^2 - d^2)}{4S}$$

The bale weight is often important in practice (Fig. 3.34). This can be calculated as follows:

$$G_B = \frac{ABLG_w}{1,000}$$

G_B = Weight of the bale [kg]
AB = Working width [m]
L = Length of the material [m]
G_w = Weight of the material [g/m^2]

3.3.2 Guiding Web Movement

Controlled regulation of web flow is another important factor in preventing non-productive times and stoppages caused by tears in release paper or the textile.

During coating, with respect to the movement of the material, the carrier material is subjected to the following influences and errors:

Dynamic errors	Static errors	Oscillation errors
Influences by the application of moisture, effects of heat, Fluctuations in web tension	Differences in thickness, misalignments in rollers	Machine vibrations, Resonance effects caused by imbalance

Modern systems come in the widest range of working widths. Depending on the edge cut, the width of the material can be calculated by the following reference value:

$$WOB = AB + 200 \text{ mm}$$

In coating systems, with one of the most common working widths of 2.0 m one has almost 10 % waste cuttings. A prime objective should be to keep this percentage as low as possible. Here, a properly designed web guiding system helps; one that corrects the movement of the carrier material with a precision of 1–2 mm.

The most common web guiding systems for textile coating systems are pivot guides and controller rollers. Pivot guides make it possible to make drastic corrections to running webs within short regulating distances. The arrangement of the rollers is swivelled along curved rails around an imaginary centre of rotation that is positioned precisely in the infeed plane. It is controlled by a pneumo-hydraulic regulation system that allows lateral placement beyond the edge of the web. Pivot guides are used for guiding short web sections. In the case of long sections, several pivotable disc rollers are used, which are usually placed after the dryer. Scanning is usually performed using an air current or optical sensor. The scanning systems work without making contact and maintain a constant temperature.

3.3.3 Material Reserves (Compensator)

The material reserves are there to ensure a continuous work process. They ensure that when a winder is replaced, enough material is still available to prevent the need for stopping the system (Johannaber 1992). If a winder runs out then a second, full winder must be available and new carrier material can be fastened to the previous material. During this time, production continues to run and the material reserves unit continues to release the carrier material it holds. The upper roller frame is run up or down by a hydraulic cylinder, or by a motorised unit, which causes the stockpile to empty or fill.

A pair of rollers are located up or downstream from the compensator (another name for the material reserves) so that the material can be fed to or taken off the unit. At speeds of 15–40 m/min a 60 m compensator (with respect to the length of material it is capable of holding) provides a grace period of 1–4 min. This period of time is sufficient for practised operating personal to sew on the new carrier material in the case of textiles, or attach it, in the case of release paper (Figs. 3.35 and 3.36).

3.3.4 Accumulator Trough

The accumulator trough is a simplified version used for accumulating textiles for the sewing-on operation. Here, the substrate is laid out in folds by a slide or a structure with a different design, in order to hold more material. When needed, the textile is fed from the accumulator trough until all folds are pulled taut.

Fig. 3.35 Compensator

Fig. 3.36 Slide trough

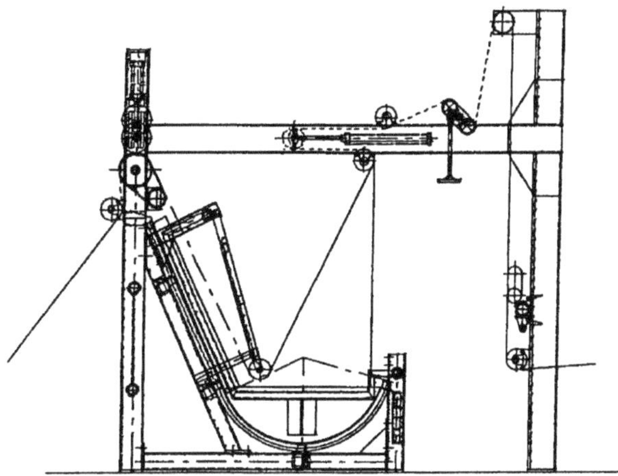

3.3.5   *Tenter Frames*

Tenter frames are used to keep the carrier material taut and wide over the necessary width. The carrier material is fed under tension into the dryer after the first coating. The tension applied prevents the coated material from shrinking at temperatures of approx. 125–165 °C. The most important part of tenter frames are the pin or clip mechanisms that hold the material on both sides to keep it flat. The tenter frames that run right through the dryer can be adjusted in width and are arranged either horizontally or vertically.

3.3.6   Drive Technology

The purpose of the drive technology is to move a sheet substrate continuously through a coating system and to roll it up with constant tension. Here, the following criteria must be satisfied:

- The drive unit must uniformly drive all of the parts of the system that are responsible for transport.
- The material must be fed through the system in such a way that the web does not run off to the sides of the transport rollers.
- It must be possible to prepare material reserves (use of compensators) in order to ensure a continuous work process.
- It must be possible to wind up the material with consistent tension, regardless of the winding diameter.

The task of the drive is to make the coating system move the way it is supposed to so that it can fulfil its specific function. The drive capacity is determined by:

1. The torque M and the angular velocity v_W for rotational movement.
   and
2. The force F and the velocity v for linear movements.

An electric drive is usually used for the drive technology because it covers a wide range of performance from only a few Watts up to 1,000 kW, and can control the speed in high threshold areas. Last but not least, the low level of noise is also an advantage.

Either direct current or alternating current is used in the coating industry, depending on the area of application. Direct current has the advantage that it offers continuously variable adjustment by a potentiometer. It is used in drive and drawing roller systems, as well as in winding and unwinding units. In contrast, alternating current drives are found in e.g. ventilators where a constant speed is required. Recently these have also being used with frequency converters and coupled PLC units to control the entire system.

In order to estimate drive capacity for these, equations are used in which the required speed of the material and the required tension are included in the calculation.

The degree of effectiveness W_{grad}, *which is also needed in the equations*, lies at 0.9 for gear mechanisms.

Drive Capacity for Drawing Rollers:

$$\text{Drive capacity } P = \frac{M \times n}{W_{grad}}$$

$$\text{Speed } n = \frac{v}{U}$$

$$\text{Torque } M = F \times r$$

$v = $ Velocity
$F = $ Force
$U = $ Roller circumference
$r = $ Roller radius

Drive Capacity of the Winding Drives:

$$\text{Winding ratio } q = \frac{D}{d}$$

$$\text{Drive capacity } P = \frac{v \times F}{W_{\text{grad}} \times 102}$$

$$\text{Req. capacity } P_{\text{req}} = P \times q$$

$D \ = $ Outer diameter
$d \ = $ Inner diameter

Reference values for the rough design or testing of the capacity to be installed are listed in Table 3.6.

3.4 Auxiliary Units

One auxiliary unit in coating systems is the cutting mechanism. It can be used to cut the final product lengthwise or crosswise. Here, it must be possible to achieve clean, largely dust-free cut edges. The cutting mechanism should be attached at either the run-out of the production line or at the inspection machinery.

There are different kinds of cutting systems, ranging from the scissor cut to crush cut system. The scissor cut system has established itself as a universal system. It consists of two intermeshed, circular knife blades and two knife shafts. The upper and the lower knife shafts are both driven by an alternating or direct current gear motor via a spur gear system. The feed towards the direction of cut of the textile to be cut must be set up so that the web meets the lower knife at the cutting point T, which is situated directly above the lower knife shaft. This means that the level on which the web is running must run parallel to the contact surface (Fig. 3.37).

The best cutting results are usually achieved with an overlap of the upper and lower knives of 3 mm. This overlap applies for materials such as imitation leather or light-weight PVC floor covering. The overlap can be adjusted depending on the material to be cut. The paper industry, for example, works with a depth of just 0.5 mm.

In the case of the crush cut previously mentioned, the cutting knife is positioned against a so-called glass-hard roller.

Table 3.6 Drive forces and capacities of motorized winders for production speeds of 2–12 m/min (approximation 1 kg = 10 N)

Maximum drive force (N)	Roller diameter (mm)	Types of material		
		Alu. foil, foil 0.10 kg/cm	Mutton cloth, PVC light fabric 0.177 kg/cm	Fabric, medium weight 0.20 kg/cm
1,000	80–600	10 kg/0.5 kW	/0.5 kW	/0.5 kW
	80–800		17.7 kg/ 0.5 kW	20 kg/1.0 kw
	100–1,000		/1.0 kW	/1.0 kW
	200–1,600		/1.5 kW	/1.5 kW
1,200	80–600	12 kg/0.5 kW	/0.5 kW	/1.0 kW
	80–800		21 kg/1.0 kW	24 kg/1.0 kW
	100–1,000		/1.0 kW	/1.0 kW
	200–1,600		/1.5 kW	/1.5 kW
1,400	80–600	14 kg/0.5 kW	/1.0 kW	/1.0 kW
	80–800		25 kg/1.0 kW	28 kg/1.0 kW
	100–1,000		/1.0 kW	/1.0 kW
	200–1,600		/1.5 kW	/1.5 kW
1,600	80–600	16 kg/0.5 kW	/1.0 kW	/1.0 kW
	80–800		28.5 kg/1.0 kW	32 kg/1.0 kW
	100–1,000		/1.0 kW	/1.0 kW
	200–1,600		/1.5 kW	/1.5 kW
1,800	80–600	18 kg/0.5 kW	/1.0 kW	/1.0 kW
	80–800		32 kg/1.0 kW	36 kg/1.5 kW
	100–1,000		/1.0 kW	/1.5 kW
	200–1,600		/1.5 kW	/1.5 kW
2,000	80–600	20 kg/0.5 kW	/1.0 kW	/1.0 kW
	80–800		35 kg/1.5 kW	40 kg/1.5 kW
	100–1,000		/1.5 kW	/1.5 kW
	200–1,600		/1.5 kW	/1.5 kW
2,200	80–600	22 kg/0.5 kW	/1.0 kW	/1.0 kW
	80–800		39 kg/1.5 kW	44 kg/1.5 kW
	100–1,000		/1.5 kW	/1.5 kW
	200–1,600		/1.5 kW	/1.5 kW

Types of material

Heavy fabric 0.40 kg/cm	Synthetic fibre, flat 0.465 kg/cm	Floor covering 0.60 kg/cm	Paper up to 100 g/m², 1.0 kg/cm	Billboard flat, two-sided 1.2 kg/cm
/1.0 kW			/1.5 kW	/2.5 kW
40 kg/1.5 kW	46.5 kg/1.0 kW	60 kg/1.0 kW	100 kg/2.0 kW	180 kg/3.5 kW
/1.5 kW			/2.0 kW	/3.5 kW
/1.5 kW			/1.5 kW	/4.5 kW
/1.5 kW		/1.0 kW	/1.5 kW	/3.0 kW
48 kg/1.5 kW	56 kg/1.0 kW	72 kg/1.5 kW	120 kg/2.0 kW	144 kg/3.5 kW
/1.5 kW		/1.5 kW	/2.0 kW	/3.0 kW
/1.5 kW		/1.0 kW	/2.0 kW	/3.0 kW
/1.5 kW	/1.0 kW		/2.0 kW	
56 kg/1.5 kW	65 kg/1.5 kW	84 kg/1.5 kW	140 kg/3.0 kW	168 kg/3.5 kW
/1.5 kW	/1.5 kW		/3.0 kW	
/2.0 kW	/1.0 kW		/2.0 kW	
/1.5 kW	/1.0 kW		/2.0 kW	/3.0 kW

Table 3.6 (Continued)

Types of material				
Heavy fabric 0.40 kg/cm	Synthetic fibre, flat 0.465 kg/cm	Floor covering 0.60 kg/cm	Paper up to 100 g/m^2, 1.0 kg/cm	Billboard flat, two-sided 1.2 kg/cm
64 kg/1.5 kW	74 kg/1.5 kW	94 kg/1.5 kW	160 kg/3.0 kW	188 kg/4.5 kW
/1.5 kW	/1.5 kW		/3.0 kW	/4.5 kW
/2.0 kW	/1.5 kW		/2.0 kW	/3.0 kW
/1.5 kW		/1.5 kW		
72 kg/2.0 kW	84 kg/1.5 kW	102 kg/2.0 kW	180 kg/3.0 kW	204 kg/5.5 kW
/2.0 kW		/2.0 kW		
/2.0 kW		/1.5 kW		
/1.5 kW		/1.5 kW		
80 kg/2.5 kW	93 kg/1.5 kW	120 kg/2.0 kW	200 kg/3.0 kW	240 kg/4.5 kW
/2.0 kW		/2.0 kW		
/2.0 kW		/2.0 kW		
/1.5 kW	/1.5 kW	/2.0 kW	/3.0 kW	
88 kg/2.0 kW	100 kg/2.0 kW	132 kg/3.0 kW	220 kg/4.0 kW	264 kg/6.5 kW
/2.0 kW	/2.0 kW	/3.0 kW	/4.0 kW	
/2.0 kW	/1.5 kW	/2.0 kW	/3.0 kW	

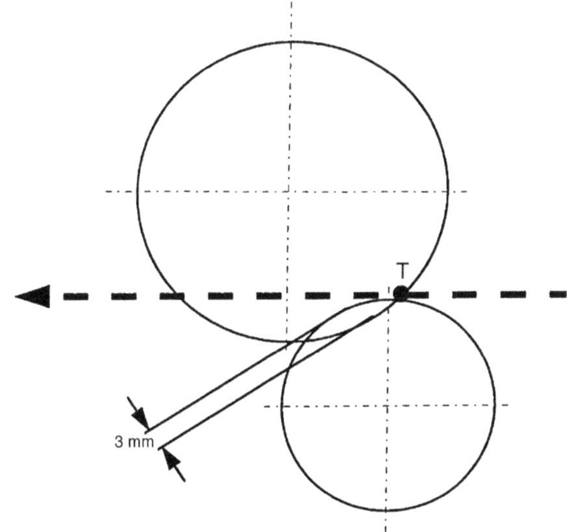

Fig. 3.37 Overlap of upper and lower knives

3.5 Drying Systems

3.5.1 Drying

In order to achieve drying, coagulation or foaming of the material it is necessary to apply heat. This heat is provided by a drying tunnel or by a heated roller. Here, the following points are particularly important:

1. Sufficient fresh air in order to disperse the solvent released by gelling and the air that contains plasticiser (Ex-protection regulations)
2. Prevention or removal of plasticizers and their precipitates which contaminate and may damage the web when they drip
3. Controlled air circulation to prevent surface-film-like effects or irregular drying and foaming of the material, which could lead to uneven thicknesses

Three basic types of drying tunnels can be differentiated (Giessmann 1982):

- Air circulation tunnel
- Heat radiation tunnel
- Air jet tunnel

This differentiation concerns the way that the hot air is transported through the tunnel as well as the way the heat is dissipated, or respectively, the way the heat is transferred to the web.

3.5.1.1   Air Circulation Tunnel

In terms of concept, the air circulation tunnel is the simplest. The heated air is blown in via blower nozzles, according to the counter-current principle; i.e. opposite to the direction the web is moving. Here, the sections can be constructed in such a way that, in an 8 or 12 m long tunnel, for example, air is blown in at two or three points, and the exhaust is located at the entrance to the section. The ratio of air circulation to air extraction is calculated in such a way that a slight vacuum is created inside the tunnel. This prevents the heated air and the plasticiser from escaping out into the room.

There is no actual zone regulation because air flowing in the lengthwise direction mixes. The ventilators and heating registers are mounted above the tunnel. The air duct is constructed to meet the needs of the specific application. The basic air circulation tunnel is sufficient for such needs as lacquering, basic PVC coating, and laminating.

3.5.1.2   Heat Radiation Tunnel

The heat radiation tunnel has a construction similar to a basic air circulation tunnel. The radiators, which are usually electric, are mounted in a tunnel with insulated ceiling, side and floor plates. The material is gelled by the radiation. A small amount of air circulation is generated by the exhaust ventilators.

Because of explosion safety regulations, the heat radiation tunnel is restricted in its applications, since the temperature of the heat from the radiators is far above permissible ignition temperatures.

Its advantages are the high intensity of the radiation, it is easy to regulate, and the reaction times are very quick. A high level of radiation intensity is needed, particularly when only short drying sections are available. The quick reaction times are also very important when starting up the production line, because this way waiting

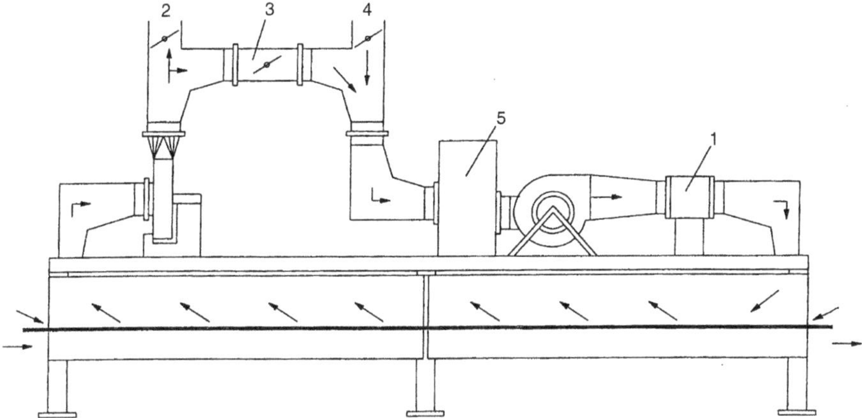

Fig. 3.38 *1* Air circulation tunnel; *2* exhaust; *3* air circulation; *4* fresh air; *5* filter

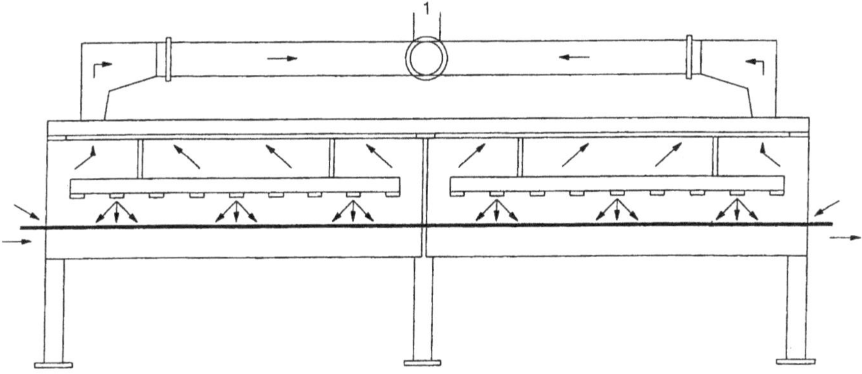

Fig. 3.39 Heat radiation tunnel (small amount of air circulation by *1* exhaust air ventilator)

times until the required heat level has been reached can be reduced to a minimum. Some manufacturers of infrared heaters guarantee immediate cooling of the radiators so that the material does not ignite if production is stopped or an emergency stop is triggered. The wavelength of infrared heaters usually lies in the range of 2.6–9.6 μm, which is equivalent to the medium and long wave ranges. The maximum temperatures that can be achieved with such heaters lies at approx. 800 °C (Krelus 1995) (Figs. 3.38 and 3.39).

An important factor in the design of the widths of such radiators is the calculation of the scattering angle. Infrared heaters assume an angle of 45° as range that is affected by the radiation. This means that the radiation has an effect on all points—especially the end points (Figs. 3.40 and 3.41).

In order to calculate the length of the radiator l_{st}, one needs the distance h_0 between the radiator and the carrier material, and the length of the carrier material that is to be subjected to the radiation l_{TM}. If the application thickness of the coating

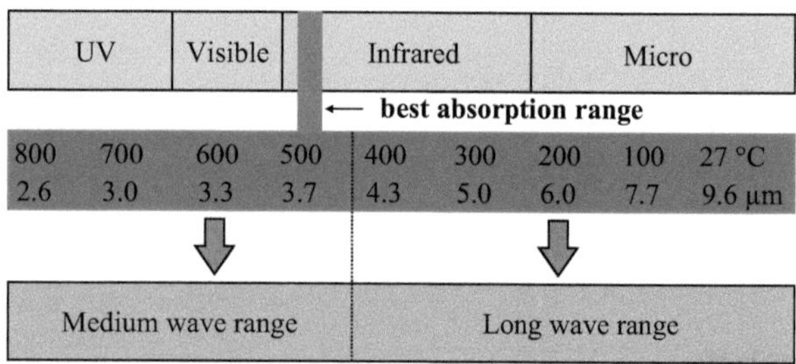

Fig. 3.40 Range of emitted wave lengths

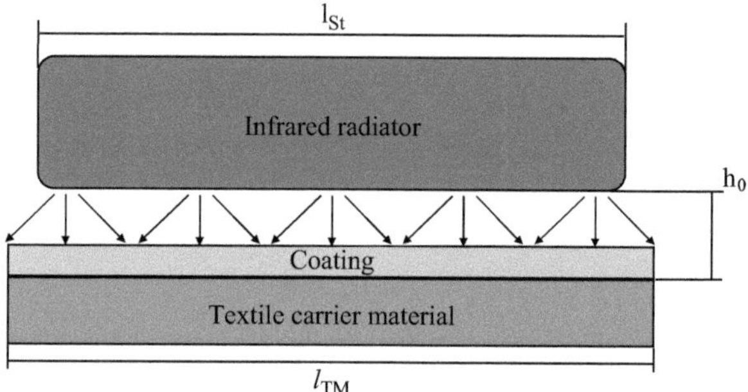

Fig. 3.41 Radiation range with respect to the working width

is very small, this can be disregarded (Krelus 1995).

$$l\mathrm{st} = l_{\mathrm{St}} = l_{\mathrm{TM}} + 2_X \times h_0.$$

If the radiators chosen are too small then there is a danger of a chimney effect on the boundary areas on the sides. The rising heat would lead to imprecise conditions on the edges, and thus, flawed results. A rule of thumb is also helpful in practice here: an additional roughly 10 cm width per side is calculated in.

3.5.1.3   Air Jet Tunnel

The most important type of tunnel is the air jet tunnel which works according to the principle of applying hot air to the material vertically through nozzles or perforated plates. The tunnel is constructed as pressure chamber; i.e. the hot circulating air is fed to the openings by a kind of diffuser. Usually the circulating air is applied to the material from above and below. Here, the ratio is approx. 2/3 from above and approx. 1/3 from below, which can be adapted as needed by a throttle valve system. The air jet tunnel is constructed in such a way that, depending on the manufacturer, either 2-, 3- or 4-m sections are placed in series to achieve the required length. Each section is equipped with its own ventilator, heating registers, and temperature controls to make it possible to regulate each individual zone. For example, in a 3-m section, the air circulates at between 5,000 and 8,000 m³/h. The height of the openings can be

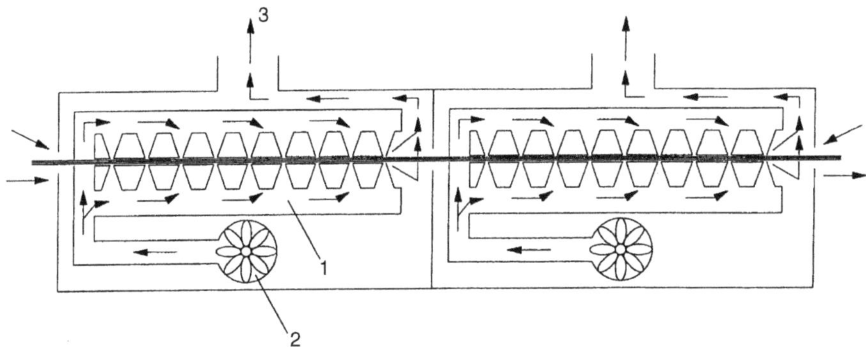

Fig. 3.42 Air jet tunnel; *1* jets; *2* air intake; *3* exhaust air

adjusted and the width of the slits is variable so that the air speed and width of the jet can both be adjusted, without having any changes to the heat conditions. The tunnels are designed for up to 250 °C, and some even up to 285 °C. The temperature must be kept constant over the entire width of the material. In practice, tunnels of between 9 and 15 m in length can be found, whereby when there are several coating heads, the final tunnel may be as long as 18 m or more. In the case of systems with one coating head, tunnel lengths of 21 m and more are not unusual, in order to achieve acceptable production speeds.

The following temperatures can be used as reference values for the above-named tunnels.

1. Coating with pastes containing foaming agents (foam PVC)   195–225 °C
2. Direct coating with PVC (compact PVC)   175–185 °C
3. Laminating, depending on the adhesive   110–140 °C
4. PU coating   90–165 °C

The temperatures to be set are dependent on the recipes used, the layout of the system, and the production speed which determine the exposure time of the coated substrates in the tunnel. This means that one can calculate the length of the tunnel based on the amount of coating to be applied, and the desired production speed (Fig. 3.42).

3.5.1.4 Heated Roller

The heated roller can be used for a very wide variety of tasks. It is used to dry the material being processed so that the residual moisture does not have a negative effect on the coat.

Apart from drying tunnels, contact drying is the most common area of application. The latter can be found in carpet backing coats, for example, where the web is dried by a heated roller and the largest possible of loops. When the composite is being dried by the heated roller after coating, something like a pressing/ironing effect takes place. Heating, which can be at temperatures of up to 185 °C, is usually performed using thermal oil, but also sometimes water if no temperatures above 135 °C are required. Steam is used only rarely today.

3.5.1.5 UV Curing Process

Drying or curing of dyes, printing inks, lacquers and adhesives in the coating industry using ultraviolet radiation (UV) is a technology whose use is increasing in many areas of application.

Ultraviolet radiation is the strongest form of radiation—invisible to the human eye—which occurs naturally in nature.

UV radiation can be divided into the following wavelength ranges:

UVC—radiation of 180–280 nm
UVB—radiation of 280–320 nm
UVA—radiation of 320–400 nm

Drying, or respectively, curing occurs as a result of the photo-chemical reaction triggered by UV radiation, whereby a solid film is generated by crosslinking or polymerisation.

Substances which can be cured by UV radiation are specially manufactured reactive liquids; i.e. no solvents are required. This means that no emissions harmful to the environment are generated during curing, besides the ozone generated by the UV system itself. However, photosensitive additives—so-called photoinitiators—are necessary.

Photoinitiators can be a problem when used with products used in the open air, because they often cause yellowing after a period of time. They are also a considerable cost factor.

Advantages of UV Curing

- Save energy (no thermal drying of substances containing solvents)
- Unnecessary to take solvent/explosion safety regulations into consideration
- Faster conversion, with correspondingly high production speeds
- The product is already fully cured directly after radiation and can be immediately finished or stored away
- Little heating of the substrate
- Good surface quality
- The application thickness is the same as the cured coat thickness

Medium pressure mercury lamps are the main kind of UV lamp used.

Because of their compact construction, UV curing or drying systems have a small footprint. No exhaust air line and usually no cooling zone are needed.

The drying area must not have elaborate heat insulation.

Because of the high degree of reactivity, UV coatings must be protected against inadvertent UV radiation. The systems must be encapsulated appropriately; lamps or windows can be covered with appropriate UV protection films, as the case may be.

UV curing is becoming increasingly important for:

- UV-curable printing inks
- Scratch and wear resistant coatings of wood, laminates, metal materials, and much more
- Printing of foils for packaging, particularly for food products
- Barrier coats

Curing Variations

Radical "Radical curing corresponds to a polymerisation reaction of alkene-group-containing molecules (unsaturated) (acrylates, vinyl ether). Special characteristics:

- The film formation is irreversible and corresponds to a polymerisation. The polymer formed is insoluble.
- Polymerisation occurs in a fraction of a second, provided the UV light source is intense enough.
- Polymerisation can only occur through the effect of UV light.
- Radical curing is inhibited by oxygen (O_2)." (Tritron 2008)

Cationic "Cationic curing corresponds to a cationic polymerisation of oxirane-group-containing molecules (epoxides) or unsaturated vinyl ethers and is triggered by cation-forming photoinitiators. Special characteristics:

- The film formation is irreversible and produces an insoluble polymer.
- Depending on the source of radiation, a cationic polymerisation is quick, but usually slower than the radical polymerisation.
- Polymerisation is triggered by UV light and, once triggered, it continues as quantitative polymerisation without any further influence of light.
- Cationic curing is not inhibited by oxygen (O_2)." (Tritron 2008)

A clear advantage of cationic curing is the insensitivity to oxygen. But levels of atmospheric moisture that are too high can have a negative effect on the curing process.

Dual In the dual-cure method two crosslinking mechanisms work in one coating medium, which causes a further curing process to take place, independent of the UV curing.

Two curing mechanisms work in one system:

- UV curing and anaerobic curing
- UV curing and thermal curing
- UV curing and moisture curing

The second curing step ensures that all areas are reached, even those that UV radiation cannot reach.

Inertisation (Oxygen Inhibition) In order to prevent atmospheric oxygen in the surroundings from reacting with the reactive double bonds in the radical coating medium (alkene-group-containing molecules), thus preventing further polymerisation on the surface of the lacquer or printing ink, the system must be inertised.

The oxygen must be removed, otherwise the lacquer or printing ink would remain sticky on the surface while deeper layers would be cured. That would mean it would be impossible to achieve product characteristics such as good scratch resistance or chemical resistance.

UV curing in an inert atmosphere is the solution.

Inert gases are gases which do not react with other chemical components in the lacquer system. The following substances are examples of what can be used as inert gases:

- Nitrogen
- Noble gases such as helium, neon, argon, krypton...
- Carbon dioxide

An IR dryer upstream from the UV dryer has a beneficial effect. On one hand this reduces the relative humidity. On the other hand a slightly higher temperature can cause a significant improvement in reaction speed during UV exposure.

Heat also accelerates the post-curing effect after drying.

3.5.1.6   Electron Beam Curing (EBC)

Electron beam curing (EBC) is considered one of the most economically modern curing methods for polymerising systems.

Electron beams are used to crosslink acrylate coatings on paper, PET, polyolefins (PE and PP), PVC, wood and metal, as well as for so-called cold curing of reactive resin systems.

Intensive research continues to develop further areas of application.

Especially on sheet products, electron beams can penetrate a monomer-containing, still liquid lacquer layer, and polymerize fully to create resistant layers. Regardless of the colour, pigment or other extenders, one achieves very well-cured coatings, as long as the lacquer coat thickness does not exceed the usable range of the electrons. The depth of penetration of electrons in the lacquer coat can be calculated and determined using more precise curves (electron crosslinking process sheet). Drying and curing takes places in a matter of seconds. Neither solvents nor photoinitiators are needed. Even heat-sensitive substrates can be dried since the polymerisation reaction triggered by electron beams does not generate any relevant heat.

Principle   The electron beam accelerator is constructed analogous to a cathode ray tube (Wikipedia—Cathode ray tube 2009 [Wikipedia—Braunsche Röhre 2009]). Electrons are emitted in a vacuum by a hot cathode, accelerated in an electric field, redirected by a magnetic field in the direction of the anode, and meet on the surface to be irradiated that is being transported through the beam curtain at different speeds. When the electrons meet the surface, the short chain molecules react to long chain polymers and thus cure the surface.

Advantages of EBC Technology

- Environmentally friendly and a technology of the future, because of the use of solvent-free coating substances
- No residual solvents
- No air or water pollution
- Good performance characteristics because of the high degree of curing of the surface
- No compromises in terms of smells or flavours because of high polymerisation
- Possible to coat thin and temperature-sensitive materials; plastic recycling possible
- No long waiting times after drying; quick finishing

Explanation of the Advantages of EBC:

- Curing of thick lacquer coatings:
  In electron beam generators the electron beams cause the curing of the lacquer, and the accelerating voltage of the generator determines the penetration depth in the lacquer. Lacquer coats of up to approx. 300 g can be reliably cured in one operation.
  In other techniques of lacquer curing the lacquer takes several hours to cure (solvent-containing) or must be applied in multiple thin layers.
- Curing of coloured lacquers:
  Using UV light to cure lacquer coats is subject to the limits of visible light. Paper as well as coloured and thick lacquer coats cannot be cured by UV light in one operation. These limits do not apply to EBC.
- Weather resistant lacquer coats:
  Electron beams generate very high crosslink density in the lacquer; i.e. virtually all adjacent lacquer molecules are crosslinked with one another.
  In the case of UV lacquers the molecules crosslink with one another only where crosslinking reactions are initiated; i.e. there where there are photoinitiators. This leads to a curing of the lacquer that resembles a spider web. Adjacent to these individual, crosslinked molecules lie many non-crosslinked or only partially crosslinked molecules. These improperly crosslinked molecules are washed out by the effects of weather such as sunlight, temperature, and rain (e.g. dulled lacquer on a car).
- Easy to clean:
  The high crosslink density leads to a very fine and dense surface. Dirt cannot become lodged in rough pores; the surface is easy to clean (Delignit).

Disadvantages of EBC Technology   The polymerisation reaction can be strongly hindered by oxygen radicals; i.e. inertisation must be carried out to achieve flawless coating results (see Sect. 3.5.1.5 UV curing).

The main disadvantage of these systems is the high acquisition costs.

That is why the use of EBC systems only makes sense in situations of large-scale production.

3.5.2 Minimum Exhaust Volume Flow

A further important factor when designing the system is the minimum exhaust volume
flow, in order to calculate dryer and ventilator capacities:

$$Q_{minL} = \frac{f \times G_{max} \times (273 + T) \times Z}{U_{ex} \times [1 - 0.0014(T - R)] \times K_{zul} \times (273 + R) \times Z}.$$

The safety factor K_{zul} for heating surface temperatures lies at 0.25, whereby the
ventilation coefficient f is calculated with 1.34.

The exhaust volume flow for these systems lies in a range of 25,000–30,000 m^3/h.

3.5.3 Temperature Control Unit

The temperature control unit is an important tool for manufacturing high quality
products, and for ensuring consistent production. It consists of settings regulators
and a regulation valve that determines the flow rate of the heated thermal oil. The
settings regulator controls the regulating valves according to a preset temperature.
The regulating valves open or close to feed the amount of thermal oil necessary for
the temperature (Samson 1994). The required tolerance between the set temperature
and the current temperature is 1 %. This means a possible temperature deviation of
less than 2 °C. The low tolerance threshold is associated mainly with the process of
foaming. It is only possible to achieve uniform gelling on the entire surface, and with
that a uniform thickness of the product, if the temperature is virtually constant. The
temperature control unit is situated both on each individual section of the furnace,
as well as on the heating drums.

3.5.4 Thermal Oil Boiler

The thermal oil boiler provides the process heat of the production system. It must be
constructed in such a way that it reaches the required operation temperature within
30–45 min after being switched on (AKB 1994). No major fluctuations of the set or
preset target temperature should occur, even if all the connected units are working
under load.

The oil boiler is designed for oil temperatures of up to 300 °C; at higher temperatures there is a danger that the thermal oil cracks and then becomes unusable. The
heated oil is fed in a closed circulation system to the pertinent heat exchangers of the
drying tunnels, or respectively, direct through secondary circulation systems to the
heated rollers. In the former case the oil radiates its process heat into the atmosphere
and heats this to the set production temperature. After the oil has been cooled this
way, it is fed back to the thermal boiler for heating again.

In the case of the heated roller mentioned in Sect. 3.5.1.4, thermal oil flows through the roller and this way warms the surface of the roller from the inside. The secondary circulation system is necessary here because the preset temperature that is required is usually much higher than the temperature that is required in the heated roller.

The second circulation system is used to feed only the heat energy to the rollers which they require, so that it is possible to achieve perfect regulation.

3.5.5 Cooling

The cooling of the web is performed by cooling rollers. Materials which may, to some extent, be in an instable condition and not capable of running through rollers, such as thick foams, are cooled in cooling tunnels. This work operation is always situated after the drying tunnels; i.e. appropriate after the application of a new layer on the carrier material and after its pre-gelling. This prevents a situation where paste gets stuck to other machine parts during its further transport.

Usually the embossing rollers are also cooled to prevent the rollers from heating up. In the case of embossing, the surface is plasticised using infrared heat. Without this work step it would be virtually impossible to achieve a uniform surface—especially the desired degree of brilliance.

3.6 Recording Data on Coating Systems for Quality Assurance Purposes

It would be impossible to achieve high quality, reproducible results without proper data recording systems for modern coating systems. In a process data recording system for a coating system, the tension, speed, drying and gelling temperatures, flow speeds in the systems, moisture, coating weight and surface dimensions, among others, are detected and systematically documented.

3.6.1 Recording Measured Values

By recording the process data, control technology within a coating system makes it possible to reproduce coating processes and to administer formulae. In particular, a high degree of reliability is demanded in the automotive supply industry, for certification of the products. Here, systems with optimum process data recording systems and automation technology are required.

A few data specifications and precise adherence to these for quality assurance purposes are discussed as follows. The consequences in cases where processes are not properly managed are described. The examples are explained with the help of

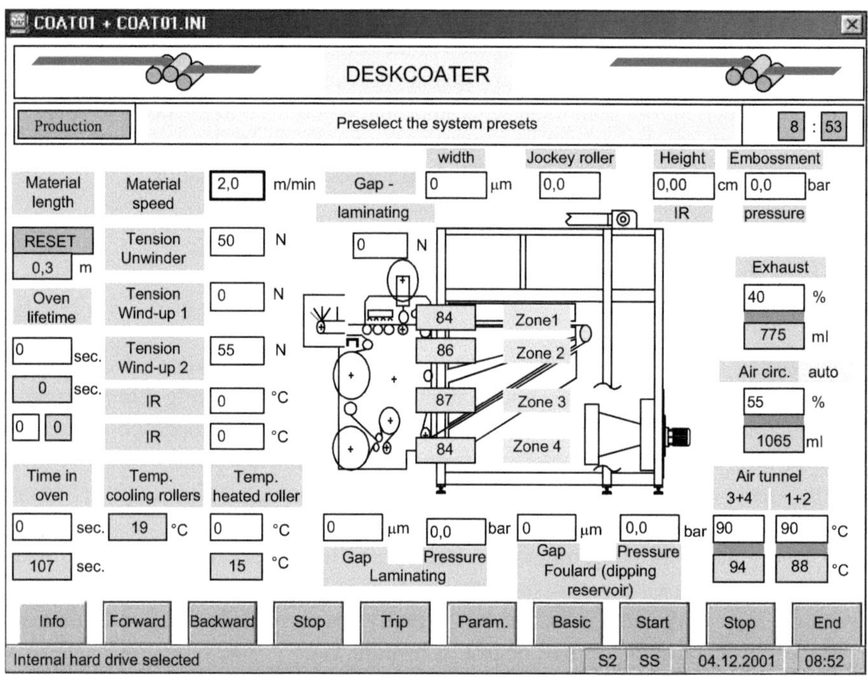

Fig. 3.43 Input screen of the Deskcoater

a multi-purpose pilot plant system: the previously mentioned Deskcoater system for coating, covering and laminating (the Coatema company). The Deskcoater is constructed as an industrial production system which was altered, however, in such a way that both unwinding as well as winding are found on the user side, and the working width is only 500 mm. The hot-air jet tunnel has two 1.5 m drying sections, each with individually adjustable temperature zones and air intakes above and below the web.

With respect to coating methods, modular replacement of the coating modules is possible (Fig. 3.43). A PLC control carries out a direct comparison between the application unit and the controls of the machine.

All machine parameters can be entered via the input screen. Included here are the tension levels of the unwinding and winding units, the production speed of the system, and the exhaust air and air circulation quantities in percent, dependent on the maximum capacity. The drying time and exposure time are defined by the material speed through the air jet tunnel.

A line diagram can display changes to the data, such as e.g. the temperatures of the individual drying zones, and the heating and cooling rollers, during the coating trials. Errors or factors influencing the temperatures can be constantly analysed.

3.6.2   Tension

The control of tension in the Deskcoater is achieved using power and speed. Using this data the computer, or respectively, the PLC controller, calculates which amount of tension is acting on the substrate from the unwinding as well as the winding up side. Both winders are motor-driven so counter control is possible when the tension changes. The motorised winding systems offer the additional advantage that, after being coated, the material can be wound up again to be given a second coating. Material which is sensitive to tension can be processed, as it were, by having the winding motor unwind directly and not brake.

The standard configuration of the control does not include functionality that determines actual values. The solution presented is one that offers precise tension control. Here it is possible to use the motor to counter changes in performance. If desired, it is also possible to control and verify current tension by using load cells.

In the case of very elastic material it is possible to adjust the speed of all motor-driven rollers, dependent on the system speed. This means that work can be performed with a lead or lag, for example. The drives are usually capable of working with a $\pm 15\,\%$ deviation to the main motor.

Maintaining precise tension during the coating process is extremely important. On the one hand, if the tension is too high or changes during coating, a lengthwise change can occur to the substrate, which can have a negative effect on the entire process as well as on the uniformity of the coating. On the other hand, if using the air knife to apply the coating, a change in tension during the coating process can cause the applied coating amount to change, which can also lead to an inconsistent coating.

3.6.3   Temperatures

A coating system contains various rollers which can be heated or cooled. Thus, one can say that the cooled roller prior to winding up the substrate is just as important as the heated roller on the embossing calender. Special effects can be achieved using temperature-controlled rollers in the case of dry or wet laminating.

A system must always have reliable temperature precision. It is essential that the various coating substances have specific temperatures and exposure times. This can be illustrated using the example of PVC coating.

Coating systems are usually divided into two drying zones, each with air from above and below. Zones 1 and 2 can be heated with a different temperature than zones 3 and 4. This is important for coating with certain types of PVC, when it is essential to start initial gelling at lower temperatures of 150 °C (zones 1 and 2) and have full gelling at 220 °C (zones 3 and 4). In the case of foam PVC, the surface is sealed in the first tunnel section so that the chemical process of foaming up can subsequently be carried out.

The temperature is measured in the system using thermocouples. Deviations during the coating process can have a big influence on the coating process. Temperatures

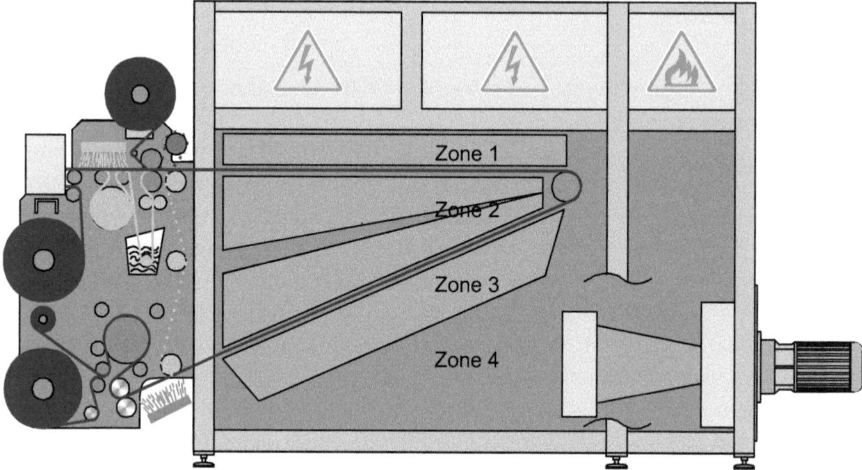

Fig. 3.44 Detailed view of the Deskcoater, with temperature zones

must be meticulously adhered to in order to provide optimum conditions for drying, crosslinking and gelling (Fig. 3.44).

For example, if the required temperature to fully gel PVC (220 °C) is not achieved, then the surface requirements with respect to abrasion, buckling resistance and similar cannot be guaranteed later when it is put to use. Furthermore, the plasticiser does not evaporate sufficiently, so surface characteristics no longer meet requirements.

When coating using chemical foams, the foaming agents cannot work properly if the drying temperature is too low. The required cell structure cannot be achieved, so the foam is instable and results in an inconsistent coat thickness. When put to use, the product does not live up to expectations.

In addition to a controlled heating process, care must also be taken to ensure sufficient cooling. Insufficient cooling before winding up can lead to the material sheets sticking together. If a coating is too warm when it is wound up it can lead to additional chemical and physical reactions which may not be particularly welcome, depending on circumstances. That is why temperatures must also be carefully maintained in the area of cooling.

3.6.4   Flow Conditions

So-called flow monitors are installed in coating systems which ascertain the flow of air and transfer the determined values to the PLC so that further action may be taken.

If the flow of air in the tunnel is too strong it can lead to problems in foam coating, but also in the case of very light materials. For example, in the case of foam coating using latex compounds, care must be taken to ensure that the air circulation—i.e. the warm air that is blown over the material—does not exceed a defined setting, otherwise the applied foam may be destroyed. Depending on the thickness of the foam applied, a change to the surface structure can also occur. Substrates which are elastic and sensitive to tension can begin to oscillate or tear. In the case of very lightweight textile webs, a combination of excessively strong air circulation during coating without a pin-chain can lead to the material flowing irregularly. This can lead to an inconsistent coating, particularly in cases of raw materials with low viscosity.

3.6.5 Coating Weight

The coating weight can be monitored online. A measurement is made before and after the coating process, and can be performed using various technologies. The difference can be used, taking into account the solids content in the coating substance, to determine the amount of coating applied.

It is possible to couple the coating weight measuring device directly with an adjustable motor-driven application head. This way the fluctuations in thickness which are ascertained can be automatically compensated by adjusting the gap of the doctor knife. The danger of an automated adjustment system like this is the amount of commotion occurring in the coating system when the gap size is constantly readjusted, and the degree of application precision deteriorates in the length of web overall. This technique is not ideal and has not established itself on the market. However, the determination of coating weight is of general importance for purposes of data recording and quality monitoring.

3.6.6 Material Moisture Level

Similar to the measurement of coating weight, there are also two measurement zones in the system to make measurements with respect to moisture: the moisture level of the substrate before coating, and the moisture level of the coated goods.

Pre-heating rollers are integrated in the system to minimise the moisture of the substrate before the coating operation. This is necessary because, in the case of certain substrates, overly high residual moisture may reduce adhesion between the polymer and the substrate. Considerable fluctuations can occur here, especially in cases where there is no air conditioning system in place where the substrates are stored. The moisture of the coated material affects the control of the drying process in terms of drying temperature and times. If material is rolled up that has not had sufficient time to dry then the web could stick together. In most cases, insufficiently dried material should also be viewed negatively in terms of the quality of the coating. The raw material for coating only sets adequately after sufficient drying, and this way achieves the required quality in terms of abrasion characteristics, surface durability, etc.

A measurement of the moisture of the material is carried out predominantly without contact, by means of microwave beams. This works according to the principle of absorption of microwaves by water. A semiconductor oscillator transmits microwave energy through the web. The unabsorbed part is received on the other side by a microwave sensor. The absorbed quantity of microwaves corresponds to the absolute moisture content of the material (see Pleva, company information). It is possible to directly couple the machine controls in such a way that when the material moisture level is too high, the machine speed can be correspondingly reduced in order to lengthen the time spent in the dryer.

Recording data in a coating system is important for safeguarding the process and optimum coating quality. In modern coating systems a direct coupling of the individual measurement systems and the machine controls with respect to temperature, material tension, moisture and other issues guarantees consistent composition of the material surface, the coating structure and the physical characteristics of the textile in the end product. The range of models varies widely. For this reason, investments with respect to the required quality should be studied carefully.

Chapter 4
Production Methods

A coating method can be described as the process of distributing a paste-like mass as uniformly as possible over the full width of a web, and fixing it in a suitable way (Schmidt 1967). In addition to transfer coating and direct coating, there is also lacquer coating which is a process used to finish semi-finished products. Other production and finishing methods which can be mentioned are printing, embossing, inspecting and final assembly. Lengthwise and crosswise cuts as well as the preparation of units for packaging fall under the heading of final assembly.

Coating systems are formed out of individual units and can have design features specific to a certain task or customer. All the units are coordinated to one another in terms of precision, work width, drive and level of automation, whereby the concept of universal use must also be included in investment planning.

Coating systems are primarily comprised of:

- An unwinding mechanism for the substrate to be coated, or the release paper.
- Material reserves with draw-in mechanism and feed-out rollers; can be a compensator or a so-called I-box
- Work platform
- Application machine, knife coating machine, roller application machine, or printing machine
- Laminating machine with substrate unwinder
- Drying and gelling tunnel with bearer drums, with or without tenter frames or transport belt, depending on the specific finished product
- Cooling unit
- Calender and embossing roller, material reserves in the form of a compensator with drawing rollers
- Pair of separating rollers to facilitate transfer coating
- Wind-up units for coated substrates and for release paper
- Drive and control elements for the system

A system planned for universal use should also be designed with explosion protection features in mind, so that it satisfies all requirements concerning water or solvent based chemicals.

A. Giessmann, *Coating Substrates and Textiles*,
DOI 10.1007/978-3-642-29160-9_4, © Springer-Verlag Berlin Heidelberg 2012

4.1 Coating Methods

The following examination of production methods assumes a universal triple head coating system. In the conception of this system special attention was paid to the needs of new investors in terms of the ability to process all common products in high quality—including ultra-thin coating—in order to enter the international market with various product ranges.

4.1.1 Transfer Coating

For transfer coating a so-called carrier medium, the release paper, is used. This is a siliconised paper which is available with different design surfaces and degrees of gloss. Normally it can be re-used ten to fifteen times; as long as it takes until the silicon surface has been worn down so much by the delamination process that the coating can no longer be separated from the surface.

In the case of the triple head knife coating machine, the release paper, arriving from the unwinding station, is fed to the first coating unit. Here, it runs through a gluing table necessary for continuous operation, and the compensator with the drawing rollers which, in the case of a roller change, automatically switches to accumulation/storage operation, and after the web has been glued, and after the emptying process, again switches over to synchronised operation. After passing through the work platform the release paper, guided along its edges by the control system, arrives at the application machine.

In this unit the paste, which is processed by roller coater or doctor knife, is uniformly applied, and then pre-gelled or pre-dried in the downstream tunnel (drying oven). After the drying temperature (approx. 155–165 °C) has cooled down in the subsequent cooling unit, the so-called foam coat is applied. This, in turn, is then pre-gelled and dried in the following tunnel at the aforementioned temperatures. This is followed by the subsequent cooling operation before the final application, the so-called adhesive coat, is carried out in the third coating machine.

In the laminating station downstream from the coating machine, the actual carrier material, a fabric, interlaced fabric or non-woven material is laminated in the wet paste and inseparably joined to the paste in the subsequent gelling and foaming process in the air jet tunnel. The temperatures here range from 185–225 °C. Following the cooling operation in which the top side and the underside must each be cooled in the cooling unit, in the separating station the carrier paper is pulled off the paste/carrier material, which has now formed into a single unit, and each is separately rolled up. Depending on how it is handled, the carrier paper can be used up to ten times or more. The rolled up product is fed to downstream machines for embossing or printing operations, if it was not already embossed by the release paper.

Calender and embossing stations are already integrated in the universal triple head coating system in order to achieve, by way of direct coating, the desired embossing effect on the surface which is specified by the embossed release paper.

Another kind of system is the steel band machine which is also used for reverse coating. These systems have a sheet of material made of continuous stainless steel band on which the coating paste is applied (Berndt 1990). This kind of coating system has not established itself in practice because of the difficulty in guiding the band and because of the temperature conditions of the bands.

4.1.2 Direct Coating

The direct coating method is used when tent canvas, clothing materials and compact imitation leathers are to be manufactured. Here, one or more coating machines are necessary, depending on the use. Coming from the unwinding station, the base material to be coated is fed to the first coating unit. The previously mentioned tenter frames are used frequently to prevent shrinking of the carrier material in the cross-wise direction. The primer or base coat is applied. After passing through the drying tunnel in which pre-gelling takes place at approx. 165 °C, and after cooling in the downstream cooling unit, the second application is applied. This, too, is pre-gelled in the air jet tunnel, and then cooled to receive the final application, the top coat. But first the actual full gelling or foaming is carried out, depending on the final product and use. The temperatures during this process lie at about 185–225 °C. The paste is applied according to the knife spreader principle (Wehlmann 1981). After the coated web has cooled, it is wound up by the winding units and, if desired, fed to appropriate systems for subsequent handling (Fig. 4.1).

Surface Treatment By treating the surface the product is given the desired appearance. Included in surface treatment are:

Grain structure:	Achieved using embossing rollers or negative embossed release paper.
Character of the surface:	Structures are achieved using inhibitors during printing, by etching the embossing rollers, or suitable flock coating.
Degree of gloss or matt finish of the surface:	Achieved with the help of various high gloss or deep matt lacquers, or by using release papers or calender rollers with various gloss effects obtained through the choice of surface or the roller body.
Colour and design:	Achieved using print rollers of different structures and combinations of different print rollers.

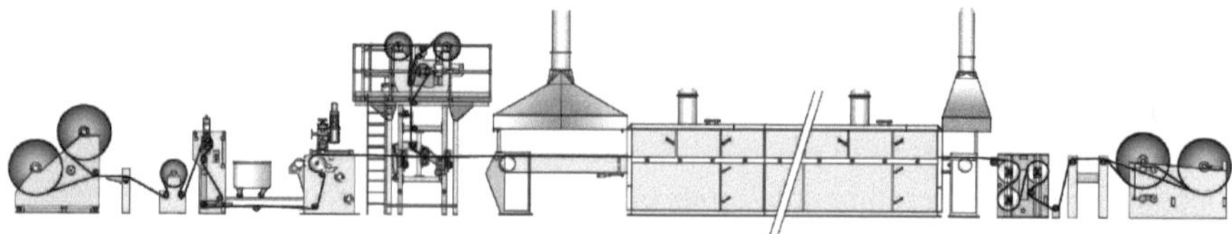

Fig. 4.1 Schematic illustration of a direct coating system

4.2 Printing

The gravure printing method is used in the coating industry for printing floor coverings, table cloths and imitation leather. In gravure printing the desired pattern is etched into the rollers. The ink is transferred to the print rollers via ink transfer rollers and the excess ink is wiped off by a doctor knife (Spitzner 1980). Up to six ink colours can be applied, depending on the type of machine. Gravure printing—also called roller printing—is a continuous printing operation (Echtermeyer 1986). Because of the flexible positioning of the print rollers, rollers positioned in sequence can be precisely coordinated to one another to achieve exact repetition. Usually roller bearings are used since they guarantee almost frictionless operation and are easy to care for and maintain (Matek 2001).

A dip roller submerged in an ink trough takes printing ink out and transfers it to the print roller situated diagonally above it. The desired image to be printed is engraved or etched in the surface of this roller so that the printing ink adheres to the recessed areas. The excess ink is wiped off by a doctor knife and flows back to the ink trough. The doctor knife has a variable design so that ink particles which have not dissolved cannot be held under the doctor knife and cause among other things, streaks. This way the particles can flow off to the side.

The ink is now transferred from the recessed areas of the engraved print roller to the web being printed. For this the printing mechanism is moved at roughly 5–10 MN towards, and pressed against, the impression roller (rubber counter roller).

The resulting print is dependent on the respective construction, and must ultimately be determined for the respective system by carrying out tests.

In the case of multi-colour printing, the web passes through several printing mechanisms with correspondingly arranged intermediate dryers, so that the subsequent print operation can be carried out without smearing the previously applied printing ink. The web is fed to the dryer after the final print operation.

The remainder of the process is similar to the coating process. After the printed web has cooled down it can be wound up.

In contrast, rotation screen printing is used for the manufacture of wallpaper. This method works with a doctor knife situated inside the printing cylinder. The ink is pumped through a tube, which simultaneously functions as doctor knife holder, into the cylinder. The metal doctor knife presses the ink through the printing cylinder (rotating screen) onto the sheet of material.

4.3   Lacquering

In order to enhance the surface of a coated substrate, be it to change the feel, to make it matt or glossy, to add additional protection against sun and/or other extraneous effects, or even to apply fashionable effects, an additional work process can be added downstream or integrated within the machine or plant: lacquering.

Lacquering is purely a surface finishing technique. Semi-finished products are given an ultra fine coating to influence the physical properties by sealing the surface, or to achieve the desired degree of gloss or matt finish. For example, this finish gives imitation leather a more pleasant feel. The treatment is also applied in the case of genuine leather. The leather industry uses silicones to give leather a softer feel or to apply a lacquer finish in various colours (Weinhold 1987).

The lacquers are applied to the coated substrate with the help of roller application systems. The application of the final lacquer is often carried out at a higher speed than the coating process itself so that in this case, separate printing-lacquering systems or separate lacquering systems are put to use.

The engraved roller is appropriate for the application of the lacquer. Engraved rollers are etched steel rollers with a very fine ribbing of 10–80 lines/cm, or also so-called thousand-point rollers (mille-point roll) which have pinprick-like depressions. These depressions can have a dull or a pointed pyramid shape, or have the shape of a ball, depending on the kind of lacquer used. Differentiation is made between viscosity, the desired amount to be applied, and between water based or solvent based consistency.

Depending on viscosity and solids content of the solution, engraved rollers allow a wet coating of 10–85 g/m^2. In the case of a 10 % solution, this corresponds to a dry coating of 1–8.5 g/m^2, which is adequate in most cases. In special cases such as with heavy tarpaulin fabrics or lorry tarpaulins, a second lacquer coating can also be applied. By adding delustering substances, glossy lacquers can be changed to matt lacquers.

Lacquers protect the printed pattern layer below them, especially against overly rapid wear. They make the surface of sticky plastics dryer, and in imitation leathers they prevent the migration of plasticisers from the coating to the surface.

4.4   Embossing

Embossing can be defined as an altering of the form of a plastic surface. Original leather structures or structures of other natural products serve as models. Since the relief is engraved as negative in the roller, the surface of the embossing roller stands out. During the embossing process the structure is pressed against a rubber roller under high pressure into the semi-finished product, which is softened by surface heat. Here, the semi-finished product is pre-plasticated by heated rollers and infrared fields, and after the embossing itself, cooled down again. Recently, it has become possible to ink the embossing roller by a following printing unit, which results in a two-tone effect. Pressure, temperature and pressing time are the most important factors for achieving lasting and good embossments (Dufke 1989).

The machines developed for embossing coated substrates in a continuous process are designed for production speeds of up to 60 m/min. Such a production system is comprised of:

- Large batch unwinder
- A pulling mechanism, if the unwinder does not have its own drive system
- Material reserves with extraction unit
- Preheater cylinder
- Infrared field, medium or short wave
- Embossing unit whereby the embossing roller and the impression roller should be water-cooled
- Print-embossing unit
- Cooling unit
- Material reserves with pulling mechanism
- Large batch winder
- Drive system

The heart of the embossing operation is the infrared field and the choice of radiators. Depending on the production speed, power lies at about 20–60 kW.

The distance to the web should be adjustable and usually preset at approx.100 mm. The gap between the rollers can also be finely adjusted and can be reset to exact positioning by a mechanism with micron precision.

The level of pressure can be set at any level and is set at up to 200,000 N. Higher pressure is possible if needed.

Many combinations of surface designs are possible and are used in the coating industry, depending on experience and know-how.

4.5 Tumbling

To achieve a special surface effect—a so-called crumpled effect—in coated substrates, they can be subjected to a tumbler treatment, a mechanical process. In an arrangement of rollers in a tunnel or in a drum, the substrate is subjected to a churning or tumbling treatment which is supported by moist, warm air and water vapour blowers. In addition, on the surface of the coated substrate, a so-called accentuated grain effect and the desired grain are achieved.

The accentuated effect looks like a coated piece of substrate dipped in warm water, and rubbed back and forth between both hands while simultaneously twisting it. The so-called creased or crash effect is achieved, in which the substrate finally resembles crumpled leather. This finishing effect also depends on the type and structure of the coating film (chemical composition of the polymer used for coating) and on the structure and the physical composition of the substrate used (web goods, knit goods or non-woven material) and the final lacquer coating, which ultimately produces the accentuated grain effect. Polyurethane coated substrates are primarily used for such a type of finishing, but also imitation PVC foam leather, which demands special conditions with respect to the composition of the foam and the adhesive strength of it on the substrate (a good adhesion).

This finishing method can take place in a discontinuous or in a continuous work process.

Calculating Gap Settings Specific gap settings are necessary in order to achieve the desired coat thickness. Calculating this depends on the specific consistency and viscosity of the paste to be worked with, as well as the speed of the material at which the substrate to be coated passes by the knife spreader.

In addition, the ratio of movement of *substrate to rotating paste* also influences the coat thickness which is formed by the gap of the doctor knife. Experience has shown that at *production speeds of 5–6 m/min.* and less, and at a *gap of the doctor knife of 1 mm, approx. 1,000 g/m²plastisol* is applied. The example calculation applies to a PVC coating, a desired coating of 850 g/m², specific weight of l:

$$v = 6\text{m/min},$$

Gap: 0.85 mm : 1.42 (specific weight) = 0.6 mm doctor knife gap.

The following apply as rough   1 mm gap = 1,000 g/m
    reference values: 0.1 mm gap = 100 g/m²
                                       0.01 mm gap = 10 g/m²
                                       0.001 mm gap = 1 g/m²

The roller gap can be calculated in a similar way.

4.6 Combined Methods

In order to achieve a surface which closely resembles genuine leather and also the effect of its feel, the following series of treatment and finishing operations has established itself in expert circles:

1. Printing
2. Lacquering
3. Printing
4. Lacquering
5. Embossing
6. Pressing (ironing)

This combination is beneficial especially because it reduces transport distances, saves manpower, and provides a continuous finishing operation. This is how so-called combined print-lacquering systems, combination systems (laminating, printing, lacquering) as well as universal printing-coating systems came into existence.

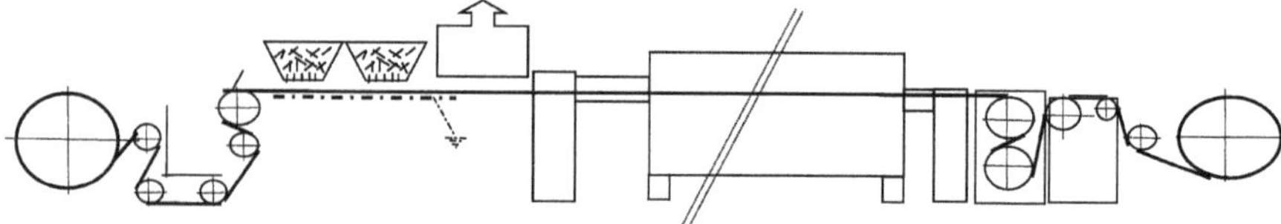

Fig. 4.2 Diagram of an electronic flock coating system. (Coatema system)

In order to produce the mentioned processes inline, and to accomplish the subsequent treatments such as e.g. pressing in high quality, it is necessary to have a high degree of process precision to achieve uniform coat thickness, a uniform application of adhesive in the case of laminated products, and uniform foaming in the case of foam PVC. Pressing is done by a heatable, highly polished roller.

Today, systems are designed and supplied that carry out inline the laminating together of multiple substrates and/or films, coating, multi-colour printing and lacquering, and—to achieve trendy effects—can even accomplish an embossing operation.

The automation of the systems is virtually limitless, but it requires increased precision of the individual units and appropriate training of the personnel operating it.

4.7   Flock Coating

A further surface effect in the case of substrates such as textiles, foils, papers and non-woven material—so-called imitation velour (velvety and plush effect)—is achieved by flock coating. Flock coating is achieved by applying short strands of textile fibres (usually cotton, polyamide or viscose) with a length of 0.5–3 mm. Depending on the effect desired and the use aimed for, shorter (almost powder-like) and longer fibres are also used.

Different adhesives with a water and/or solvent base are used for cotton, spun rayon, jute, and synthetic fibres in stack lengths between 1 and 10 mm (Fig. 4.2).

Firstly, the substrate to be flock coated is coated with a suitable adhesive using a spreading unit (knife against roller or roller application machine). Using a vibrating mechanism, the flock is dispersed (flocked) through an appropriate sieve onto the substrate either on the full surface, only points, or in a design (engraved sieves, print design) and electrostatically charged, which causes the fibres to become fixed in the adhesive, ordered and to stand erect.

The drying and gelling of the adhesive is performed in the drying and gelling tunnel after the flocking units, whereby the tunnel temperatures can range between 150 and 200 °C, depending on the adhesive. Adhesive coatings in the range of 80–150 g/m^2 have been used for full surface flock coatings.

The adhesives require excellent adhesive strength, flexibility and elasticity; and depending on the intended use of the flock coated substrate, also good water and solvent resistance, as well as wash, cooking and petroleum ether resistance.

In the case of oriented flock coating—as opposed to matted pile in which the fibres are arranged in a disorderly way—the electrostatic method is used.

In the most frequently used variation, coating is done in an electrostatic field at a direct current of roughly 100 kV. The individual fibres straighten up and stand erect in the adhesive. After drying and cooling a brush mechanism sweeps away the loose or poorly glued fibres.

The layout of the system is as follows:

1. Application of the adhesive in select quality, using a spreading unit or a roller application unit
2. Application of the flock fibres in an electrostatic direct current field, using a vibration sieve, either over the full surface or as a print design
3. Drying and gelling of the flocked web in the drying tunnel downstream from the flock coating unit (the tunnel lengths here are dependent on the desired production speeds; tunnel lengths of 18 m, 21 m and more are quite common)
4. Cooling to room temperature of the flocked web
5. Removal of the excess fibres (flocks) by a brush and vacuum system
6. Winding up of the web. Here, an edge cut and web control system can be integrated.

4.8 Laminating

If multiple substrates or layers are joined together inside two layers so that one product results in the end, this is called a lamination. The joining together of multiple layers is called a laminate. In today's parlance, however, these fine points are no longer subject to differentiation.

A few application examples are provided in order to explain the term "lamination".

In the production of coated substrates, one differentiates between direct coating (e.g. tarpaulin materials, table cloths, camping and sailing materials) and indirect coating which is used for classic, soft imitation leather (outer material for shoes, bag-making).

In the manufacture of imitation leather the top coat—a PUR or PVC layer—is applied to the carrier paper (release paper) by a knife spreader. A layer of PVC foam is applied onto this top coat. Now, the substrate to be coated is placed in this or a third, applied adhesive layer (may also be a PVC layer) and, depending on the desired effect it is held by pressure or a gap between rollers, and pressed into the wet paste.

After passing through the drying and gelling tunnel, the layers of coating are now permanently fixed to the substrate placed on them (laminated), and can be separated from the carrier paper (release paper) before winding up.

Fig. 4.3 Pressure laminating

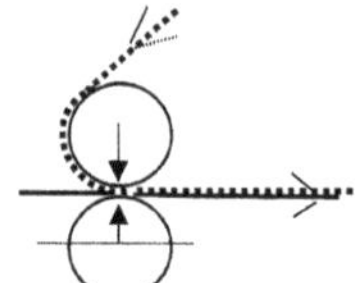

Fig. 4.4 Gap laminating

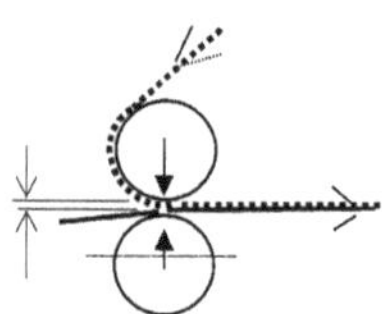

Fig. 4.5 Air laminating

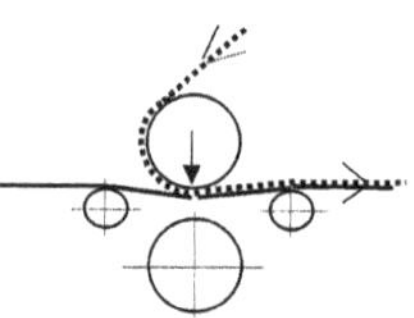

The laminating unit is essentially comprised of a chrome-plated or Teflon-coated laminating roller that works with high precision and concentricity. This laminating roller, for which there is also a coolable version for hot-melt systems, is pressed against an impression roller by pneumatics or hydraulics. The impression roller is usually a rubber roller, whereby a rubber between 65 and 87° shore hardness is chosen. It is advantageous if it is resistant to solvents, because one does not always know in advance which chemicals will be used. The laminating roller can be set up either to apply pressure or work with a gap between it and the impression roller.

A third variation is so-called air laminating that works using support rollers situated before and after the impression roller, which are also adjustable in terms of height. With this third technique, for example, two tension-sensitive substrates can be lightly laminated one above the other. They touch one another only by their own weight and by the tension factor created by the drawing of the material. The method is preferred for laminating roughened fabrics, so as to avoid compressing the roughened surface, but still to achieve good adhesion.

Figures 4.3, 4.4 and 4.5 show schematic illustrations of the three laminating methods.

In special cases the entire laminating unit is set up to be moveable in the lengthwise direction in order to adapt to customer needs and the use of different chemical substances. This moveability in the lengthwise direction means that the distance, and thus the evaporation zone, between the actual coating and the laminating can be individually adapted to the working life of the chemicals applied (Fig. 4.6).

On the one hand, this has the advantage that when working with solvent-based adhesives, covering occurs immediately and release of solvent is limited. On the other hand, the distance to the laminating gap can be enlarged in such a way that an air-drying effect is initiated and a slight skin forms before laminating, and a strengthening of the adhesive takes place which can improve adhesion overall.

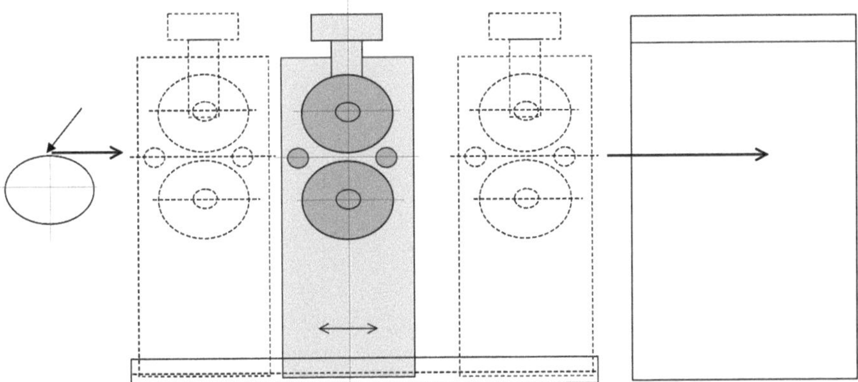

Fig. 4.6 Moveable laminating setup

Fig. 4.7 OCS dry laminating

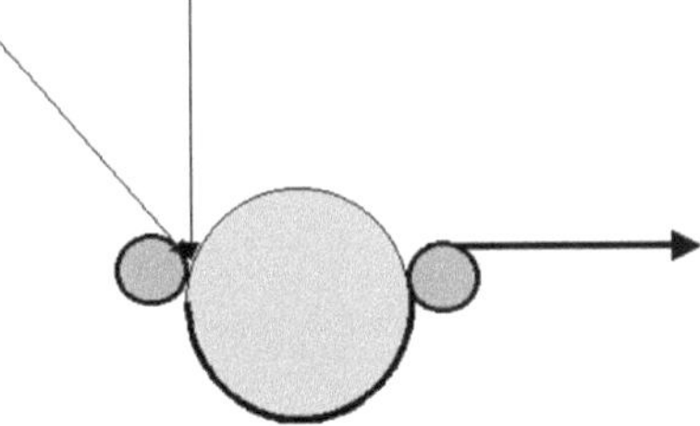

There are three different methods of laminating: *dry laminating*, *wet laminating*, and *flame laminating*.

4.8.1   Dry Laminating

Dry laminating is used when open-pored substrates must be laminated. This kind of lamination prevents the penetration of the adhesive layer into the substrate. This has an important influence on the feel.

The laminating media are formed into one layer and subsequently dried. The reactive adhesive layer is laminated to the web using a second source of heat and by melting, and consolidated under pressure. Two rollers are used for this. One is situated at the entrance to the heated roller, the other at the exit. The subsequent cooling operation completes the process, which is also known as the One Coat System (OCS). When running into the system the materials are fixed with a rubberised roller and joined together by temperature and tension. The chrome-plated exit roller is positioned as pressure roller (Fig. 4.7).

If an adhesive film is used as intermediate layer, then it is necessary to have constant pressure during the joining process, which can be achieved with a pressure cloth (Fig. 4.8).

Fig. 4.8 Belt drum dryer

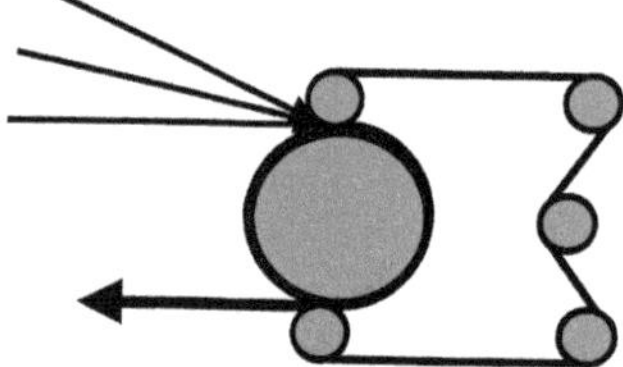

4.8.2 Wet Laminating

In the case of wet laminating, a paste medium is applied by an application system to one of the webs to be joined, and then an additional layer is added. The joining process is supported by pressure and/or heat during the laminating process.

The full curing of the application subsequently takes place, and the final thermal operation is the cooling. Dependent on the substrate it is usually cooled down to 20–25 °C, whereby for aluminium coils 40 °C is also common.

4.8.3 Flame Laminating

In the case of flame laminating, an open flame is used to melt the surface of the substrate. Usually this is a PU foam which must be joined to a covering fabric. The gas burners used here allow temperatures of approx. 200–250 °C and, throughout the melting operation, speeds of up to 40 m/min. In this kind of lamination, however, a melting off of the foam film thickness of between 0.2 and 0.5 mm is to be expected.

This technique is increasingly moving to the sidelines for ecological reasons, so many manufacturers no longer use it nowadays. Intensive work is in progress to replace this form of laminating with other technologies. Currently hot-melt applications, powder coating application, and the use of melting films are being studied. However, a satisfactory solution has not yet been found with respect to aspects including lower production and manufacturing costs.

4.9 Final Assembly and Inspection

Inspection of the substrates is necessary in the finishing process, be it for coating, laminating or printing, because these are usually not delivered in the required lengths.

This results in the necessity to sew together the delivered lengths in order to guarantee a continuous coating process. Flawed areas must be marked and removed particularly in the case of imported substrates because they could lead to interruptions or waste during the finishing process.

These work operations are necessary in the coating industry in order to adapt the finished goods to the customer's specific needs. This work is now largely automated, so that rolls suitable for sales in terms of length and width can be automatically packaged and labelled. Figures 4.9 and 4.10 show inspection machinery for wound

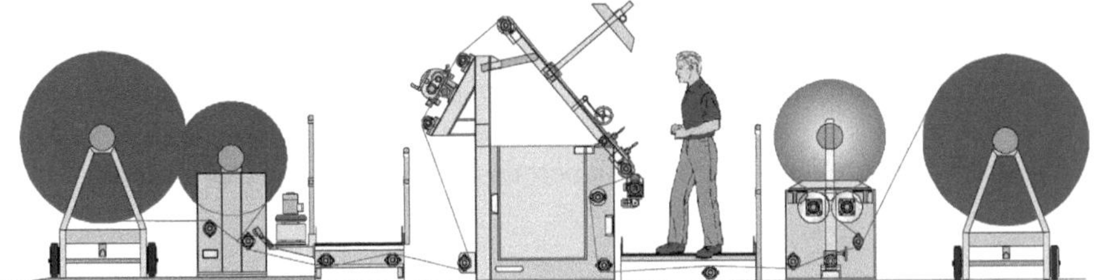

Fig. 4.9 Layout of a substrate rewinding, inspection and measuring machine

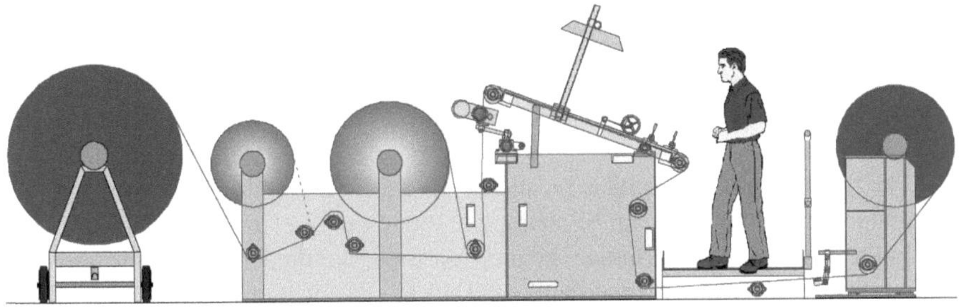

Fig. 4.10 Layout of a paper rewinding and inspection machine

goods and coated substrates, as well as for the rewinding, cleaning and inspection of release paper.

In both versions different winding systems are integrated in one system so that work can be performed from large batch winder to large batch winder, but also adapted to various length requirements. This allows even small meterages for finishing to be joined together to the desired lengths (sewing, fusing, tacking).

This means that the end product must be cut to the desired width and length. Attention is also paid here to the proper labelling and the necessary diameter of the cardboard jacket. Furthermore, during cutting the material is subjected to a quality check while it passes over the inspection table. Only the surface quality is checked during this operation. The physical and chemical properties are tested in a separate laboratory. The release paper required for indirect coating must also be inspected, and if necessary cleaned and repaired. The work is concluded with the packaging into the batch sizes expected by the customer.

In the paper rewinding and inspection machine additional winders and guide rollers are integrated to also carry out a separation of release paper and coated substrate. Experience has shown that inline separation is not always favourable, and further, additional aging of the coating is more beneficial. The chemical products used play a role here.

Appropriate means of transportation are necessary to bring the wound goods (substrates of all kinds) to the individual machinery. Not all production halls have a hall crane set up that can serve the winders of the coating and laminating machinery.

The large batch winder serves as suitable winder and transport unit here, because it is mobile and equipped with a drive key and material brake.

Another typical means of transport for web substrates is the bale transport wagon. However, it must be mentioned here that the weight of the bale causes a certain amount of load pressure that may not permissible for every kind of material.

4.10 Reference Formulae

In addition to the production methods presented, the formulae or the makeup of the chemicals used are also essential parts of production. Depending on the desired product characteristics, the chemicals are selected and combined in specific mixing ratios. Mixing is performed using the equipment described in Chap. 5 (Tables 4.1, 4.2 and 4.3).

The three reference formulae presented have established themselves as principle formulae for coating, and can be adapted to the respective range of products.

In the description of formulae the unit "parts per hundred" has become accepted. This means that the main component—e.g. PVC—is given in 100 parts, and the rest of the additives always refer to 100ths of the main component.

Table 4.1 Reference formulae for tarpaulin materials

Base formula	(kg/m^2)
Top coat	0.3
Foam coat	0.2
Total weight	0.5

Composition of reference formula

Raw material	Top coat	(kg/m^2)	Foam coat	(kg/m^2)	Total weight (kg/m^2)
S374MB	100.0	0.159151	100.0	0.10204	261.19
DINP	75.0	0.119363	75.0	0.07653	195.89
Interstab M767	2.5	0.003978			3.98
Interstab 802			2.0	0.00204	2.04
Estabex 2381	3.0	0.004774	3.0	0.003061	7.84
Interside ABF2	2.0	0.003183	2.0	0.00204	5.22
TiO$_2$ CL2220	6.0	0.009549	6.0	0.006122	15.67
Vulcabond VP			8.0	0.008163	8.16
Total	188.5	0.3	196.0	0.2	

Table 4.2 Reference formulae for imitation leather (upholstery)

Base formula	(kg/m^2)
Top coat	0.11
Foam coat	0.23
Textile	0.1
Total weight	0.44

Composition of reference formula

Raw material	Top coat	(kg/m^2)	Foam coat	(kg/m^2)	Total weight (kg/m)
S374MB	100.0	0.060439	100.0	0.115577	176.02
DINP	70.0	0.042307	70.0	0.080904	123.21
Genitron AC4			2.0	0.002311	2.31
Interstab M767	2.0	0.001208			1.21
Estabex 2307	3.0	0.001813			1.81
Sicostab M62			2.0	0.002311	2.31
Durcal 5			20.0	0.023115	2.31
TiO$_2$ CL2220	4.0	0.002417	2.0	0.002311	4.73
White spirit	3.0	0.001813	3.0	0.003467	5.28
Total	182.0	0.11	199.0	0.23	

Table 4.3 Reference formulae for rain clothing

Base formula	(kg/m^2)				
Top coat	0.18				
Foam coat	0.18				
Textile	0.09				

Composition of reference formula					
Raw material	Top coat	(kg/m^2)	Foam coat	(kg/m^2)	Total weight (kg/m^2)
S372 LD	100.0	0.108892	100.0	0.108892	217.79
DINP	60.0	0.065335	60.0	0.065335	130.67
Byk LPD 6296	0.3	0.000326	0.3	0.00326	0.65
Interstab M767	2.0	0.002177	2.0	0.002177	4.36
Estabex 2307	3.0	0.003266	3.0	0.003266	6.53
Total	188.5	0.3	196.0	0.2	

Chapter 5
Paste Preparation

Materials made of polyvinylchloride (PVC) or polyurethane (PU), acrylates or silicones as well as a series of additives are necessary ingredients in order to be capable of equipping coating systems individually and flexibly. These materials are mixed by the respective manufacturers themselves before processing. Figure 5.1 provides an overview of the equipment necessary for this. The chemicals can also be fed to the mixing machines by automatically controlled or computer controlled scales and silo units.

Five different pieces of equipment are necessary for preparation of the pastes:

- Vacuum sinus dissolver
- Vacuum filter
- Three-roller mill
- Wall-mounted high-speed mixer
- Solvent mixer

5.1 Vacuum Sinus Dissolver

Vacuum sinus dissolvers are mixers with a low- to high-speed vertical agitator.

The agitator is constructed in such a way that its kinetic energy is translated to a great extent into mixing and combination energy and not frictional heat (Finkmann 1986).

In the dissolver, PVC pastes are combined with plasticisers, kickers (foaming agents), stabilising agents, extenders, and master batches (colours), and dispersed and worked into a paste until the desired viscosity has been achieved. The paste is simultaneously placed in a vacuum to remove trapped air.

An edge wiper is recommended, depending on the viscosity. This is a knife which rotates counter to the motion and wipes the paste from the edge and feeds it to the dissolver plate. This ensures that mixing is consistent and homogeneous (Fig. 5.2).

A. Giessmann, *Coating Substrates and Textiles,*
DOI 10.1007/978-3-642-29160-9_5, © Springer-Verlag Berlin Heidelberg 2012

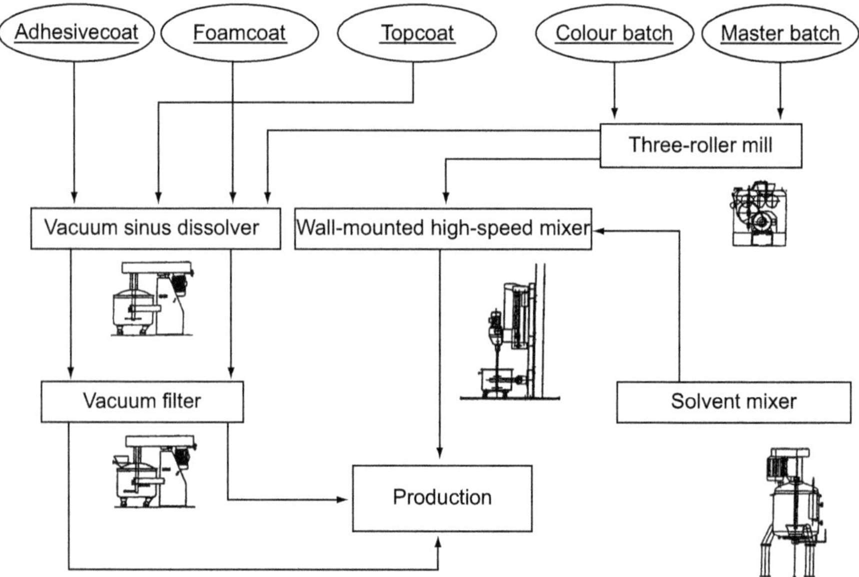

Fig. 5.1 Scheme of paste preparation. (Coatema 1990)

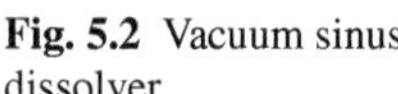

Fig. 5.2 Vacuum sinus dissolver

5.2 Vacuum Filter

The vacuum filter is used to additionally filter and subject the top coat in PVC coating to a final vacuum, in order to filter out even the most minute of impurities or concentrations of pigment which were not homogenised. It has a construction similar to the vacuum sinus dissolver, and an additional filter which is fastened in a

Fig. 5.3 Vacuum filter

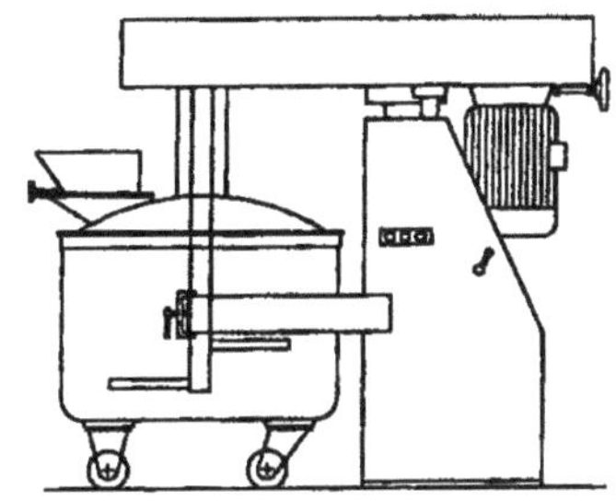

Fig. 5.4 Three-roller mill

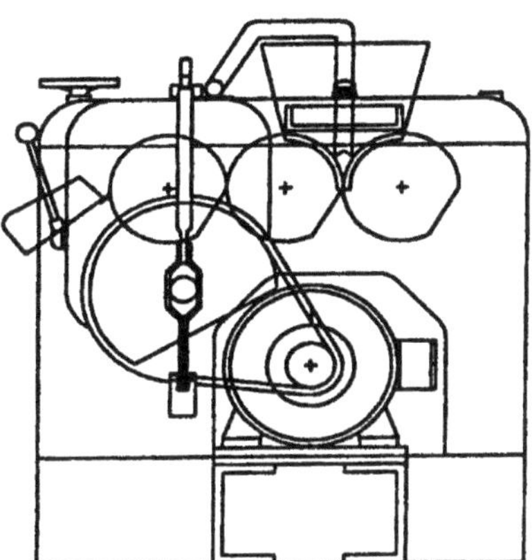

quick-change sieve frame. Instead of the rotating shaft with agitator disc, an agitator arm moves the coating substance slowly from the bottom to the top and thus supports the vacuum operation (Fig. 5.3).

5.3   Three-Roller Mill

The three-roller mill is used to grind and homogenise master batches (colours) and extenders before they are added to the PVC in the dissolver (Werner 1988). This is necessary because colour pigments and extenders are available in different grades of fineness on the market, but extremely fine media are necessary when carrying out fine coating. The master batch is fed to the outer roller and, because of the difference in speed, transferred through the middle to the third roller as film. The rubbing/abrading or milling effect is generated by the contact pressure (Fig. 5.4).

5.4   Wall-Mounted High-Speed Mixer

The wall-mounted high-speed mixer is used to dye dissolved PU pastes and—where necessary—make adaptations to viscosity (Niemann 1991). The mixer works at very high speeds in the range of 600–6,000 min^{-1} (Fig. 5.5).

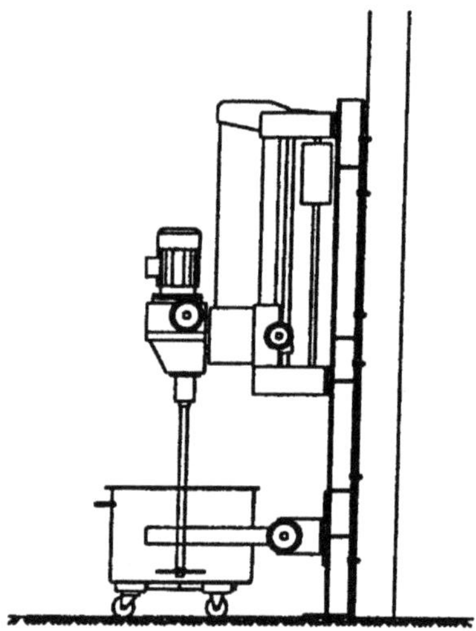

Fig. 5.5 Wall-mounted high-speed mixer

5.5 Solvent Mixer

PU granulate is placed in the solvent mixer. This is transformed into a spreadable paste by the addition of solvents such as methyl ethyl ketone, dimethyl formamid, totuol and others. A slowly running agitator mechanism ensures a homogenous dissolving process, which can take up to a few hours. The heat generated by the dissolving process is dissipated by the cooling effect of container's walls. This is extremely important in order to prevent gelling and vaporisation of solvents.

The PU solution is normally supplied in a proportion of 20–30 % solids content for the coating process. It can be coloured according to the customer's wishes by using the wall-mounted high-speed mixer (Fig. 5.6).

5.6 Plastisol

5.6.1 Preparations

The mixer types were described initially for purposes of understanding the machinery. Information will now be given on how to use these different types.

In order to simplify, let us divide the mixer types into two categories: mixers with high intensity, and mixers with medium or low intensity.

The choice of a specific kind of mixer depends mainly on the end product to be manufactured. Mixers with medium or lower intensity are used to manufacture pastes with high viscosity. Such mixers should be used for industrially manufactured mastic resins. In a coating plant where plastisols have low viscosity at high shear rates, and a low to high viscosity at low shear rates, depending on the coating, a mixer with

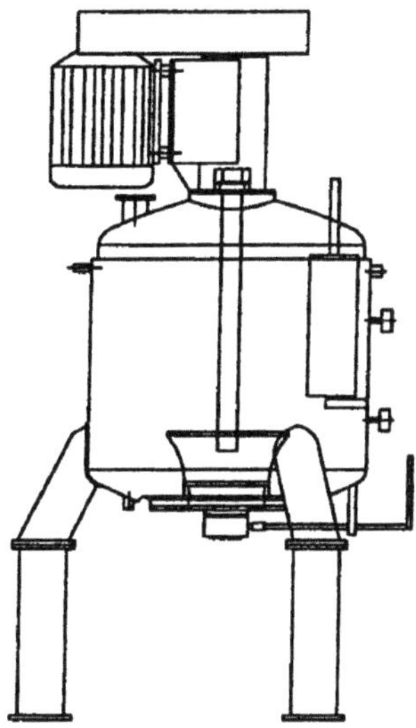

Fig. 5.6 Solvent mixer

high intensity should be used. Naturally, this is generally a matter of opinion because other aspects must also be considered. The main purpose of mixing is to reduce the degree of agglomeration present in the dry powder while dispersing it in the liquid medium. Moreover, this should be performed with little generation of heat in the paste in order to prevent premature swelling of the resin. Safe mixing temperatures can vary considerably in the different kinds of formulae.

5.6.2   Using a High Speed Mixer

When working with a special agitator blade at high speeds a vortex forms, and the particles deagglomerate because of the turbulence and shear forces. High speed mixers are generally equipped with a vacuum mechanism and sometimes also have a scraper on the edge. During the mixing process in a high speed mixer, a large amount of air is pumped into the paste which must be removed before the coating operation in order to prevent bubbles or craters. The paste can be vacuum treated in the mixer itself or in parallel during vacuum filtering.

With respect to viscosity, in literature values of 1–30, 1–25, 1–15 Pa s are specified as a maximum. Personally, if no other specific requirements are made, I would say that the lowest viscosity at high rates of shear should be the aim of the formula mixer. It is clear that as general rule, highly viscous materials and high shear rates result in a high rise in heat. Temperatures of 30–40 °C are not uncommon if you are in Africa or Asia. The formulator should work more carefully in such conditions. The surrounding temperature is one thing. But even worse are high temperature differences caused by shear rates that are too high, or insufficiencies in the formulae. Thus, the following must be observed:

- Surrounding temperature
- Temperature differences during the mixing operation
- Mixer efficiency vs. revolutions per minute; strive for shorter cycles without generation of heat
- Kind of formula: high or low plastication,
  Presence or lack of plasticisers with high solubility, extenders, pigments

Viewed from the standpoint of the equipment, double-walled mixing containers can be used to some extent to prevent the generation of heat, to achieve better control of viscosity, and to keep the remaining formula constant using temperature control.

Mixing Process First the liquids are poured into such mixers and then homogenised, and then the solid substances are added (not everything at once; dependent on the capacity of the container). The order of addition of each solid substance depends on the formula. PVC is generally added first. After this, the order and the times are specified by the formulator. The procedure must always be adhered to. Mixing times with these mixers usually varies between 15 and 30 min. After all ingredients have been combined and are in a paste form, it is important to remove all solid materials from the container walls and the agitator blade. When the paste is in a finely liquid form then the final mixing can begin. In the meantime the vacuum pump has been switched on to ensure appropriate de-airing. When trapped air is released in a vacuum the paste can accumulate as foam. In a normal formula, the vortex which is formed by the propeller should retain the foam inside the container, by breaking it up to some extent. In the case of some critical formulae, care must be taken that the foam does not rise too high and get sucked through the air discharge. If this occurs, a less powerful vacuum should be used.

Other mixer types such as paddle or ribbon mixers are also used. But they are less efficient and the cycle times are higher. In these cases, overheating is seldom a problem.

5.6.3 Low Speed Mixers

For pastes or dough-like mixtures in which solid additives must be dispersed, low speed mixers are used. First, either the liquid or solid substances are filled, depending on the quantity (the solids first in the case of small quantities). But in both cases the liquids must be added in sufficient quantities in order to form a stiff paste (a mastic phase). Then the rest is usually added in a way to maintain the mastic phase. The remaining liquids are then slowly added to create the finished paste. It is mixed further in order to ensure good homogeneity. The development of the mastic phase is very important for avoiding the production of clumpy pastes.

5.6.4 De-Airing

Trapped air can be minimised through the correct choice of additives. Highly soluble plasticisers, for example, tend to increase viscosity and impair the good de-airing characteristics. The use of some PVC resins sometimes results in poor de-airing. It is revealing to observe the surface tension of plasticisers and to test the de-airing characteristics of PVC resins.

Nevertheless, large quantities of air are pumped into the plastisol during the mixing process, especially in the case of high speed mixers. Air is also added to the plastisol through the trapped air from the surface of the powder additives (such as pigments and extenders). In order to achieve good results, in the past the plastisols were left for 24 h in a ripening process. In such cases, in pastes with low viscosity, it is possible that the air escapes on its own. This takes time. In fact most plants which are geared to production use the pastes practically directly after preparation. The use of additives can also be helpful: partial replacement of primary plasticisers by DOA when other necessary properties such as the ability to bend at low temperatures are to be improved and are economically affordable. The use of a substance to reduce viscosity, and the use of additives that reduce surface tension generally help the air to escape more quickly. Byk 3155 is a helpful additive.

Today, most mixers are equipped for vacuum operation. Vacuum operation is carried out dynamically during the mixing operation. De-airing is recommended up to a compression of 5–6 mm. Unwanted moisture can also be removed very efficiently.

In large plants an additional piece of equipment can be used. The plastisol is fed into the centre of a turning disc, rotating in a vacuum. The centrifugal forces cause the plastisol to form a thin layer on the surface of the disc. This is highly efficient. In rare cases vacuum filtration machines are used in combination with mixers that do not work in a vacuum. Under the right conditions a three-roller mill can also be used if none of these machines are available on site. It is more suitable for heavier pastes and can also capture additional air. It is necessary to differentiate between poor air release and air intake (As described in Chap. 7).

5.6.5 Filtration

When the manual addition of solids such as PVC and extenders in mixing containers has been concluded, bits of paper or other materials may have been unintentionally added. That is why the plastisol must be filtered to avoid this type of coating problem. This operation can continue in a vacuum filter. Essentially, a vacuum filter is constructed like a dissolver—usually without the edge wiper—but with an upstream filter situated in the cover. The plastisol is sucked through the filter by the vacuum. The plastisol in containers is mixed under a vacuum to remove all trapped air.

Chapter 6
Substrate Coating

During the 1980s and 1990s the focus of the coating industry was placed on the development of new technologies and application systems, as well as on the integration of additional units such as corona or plasma pre-treatments, UV radiation or drying using IR and NIR fields. Efforts increased to develop hot joining substances, so-called hot-melts or melted glues, and the necessary application technologies for the growing laminating and coating industry, as well as film coating. New application machinery had to be developed for very special applications such as nano, film and extrusion coating. The rapid response to the needs of the market lead to multifunctional coating heads, which allow the textile finishing industry a high degree of variability and flexibility. The desire in the industry for laboratory and pilot plant systems was satisfied by the development of laboratory systems which made it possible to transfer new coating solutions directly to a large production line.

In recent years, the focus in the area of textile coating was on the development of new raw materials. Examples of this are hydrophobising with simultaneous water vapour permeability, as well as the provision of specific characteristics such as electromagnetic shielding, electrical conductivity, and insulation, especially in the area of protective textiles. A hot topic in the area of textile coating is the development of water-based systems, with the aim of environmental protection and the replacement of solvent-based systems.

Today, coating is a basic requirement of a large proportion of technical textiles, which would not be capable of satisfying the strict requirements on functionality and standards without the application of special raw materials. Despite a stagnating textile market, the trend towards high quality coatings is rising unabatedly. Solutions for special coatings and increasingly faster product development demand a high degree of cooperation between machinery manufacturers and raw material suppliers in the chemicals industry.

6.1 Properties of the End Product

In order to be capable of influencing the properties of the end product, it is necessary to consider all of the components involved in the coating. The profile of the end

A. Giessmann, *Coating Substrates and Textiles,*
DOI 10.1007/978-3-642-29160-9_6, © Springer-Verlag Berlin Heidelberg 2012

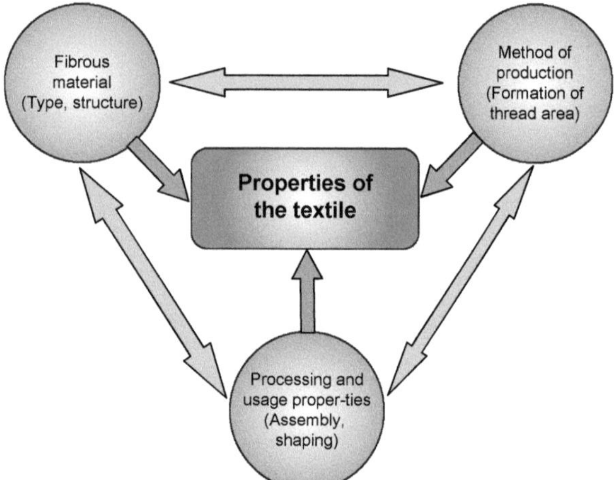

Fig. 6.1 Properties and composition of the textile carrier material

product is dependent on the fibrous raw material, the yarn used, the type of surface formation process, and the surface finish itself. The three main components are:

- Properties and composition of the textile carrier material
- Properties and composition of the coating medium
- Cohesion and composition of the composite material

6.1.1   Carrier Material

In terms of its properties the textile carrier material is comprised of the base raw material of the fibres, the method of production, and the type of finishing used.

6.1.1.1   Fibrous Material

As base material, the textile fibrous material is a structure that is limited in length (spinnable fibre) or practically endless and linear (filament). All fibrous materials have common properties: a length that is larger than its cross section, as well as good strength and the ability to bend (Fig. 6.1).

The fibrous material consists of many thousand like-sized molecules; the macromolecules. Depending on how it was produced, differentiation is made between the following:

- Natural fibres
- Chemical fibres made of natural raw materials
- Chemical fibres made of synthetic raw materials

In addition to the finishing process, the starting product for the fibrous material brings with it the base properties of the textile surface. Tables 6.1 and 6.2 list advantages and disadvantages of chemical fibres and inorganic fibres (Table 6.3).

Table 6.1 Properties of chemical fibres

Filament or yarn raw material	Advantages	Disadvantages
PE	High strength Very high E-modulus No absorption of moisture Good chemical resistance Low density	Poor temperature stability Low melting range Highly combustible
PP	Good abrasion resistance Low density No absorption of moisture Good tensile strength Low heat conduction Low chargeability	Poor temperature stability Non UV resistant Low melting range Poor dying performance
PTFE	Very good chemical resistance No corrosion whatever Good temperature stability Good light stability Good weather resistance Good scrub resistance	High price High thermal expansion coefficient Low strength High specific weight
PES	Low absorption of moisture Good dimensional stability High elasticity Good chemical resistance Good scrub resistance	Poor lye resistance
PA	High elasticity Good chemical resistance High strength	Highly combustible High photosensitivity Fibres damaged easily by thermo-oxidation
PAN	Good chemical resistance Good light stability Good strength	High fibre swelling Poor dying performance Highly combustible

Table 6.2 Properties of inorganic fibres

Inorganic fibres	Advantages	Disadvantages
Glass	High strength Good chemical resistance Good temperature stability Good tensile strength Fully elastic elongation	Low transverse strength
Carbon	High strength Good chemical resistance Good temperature stability High E-modulus Electrically conductive Low material fatigue	Very expensive Low scrub resistance

Table 6.3 Properties of natural fibres

Natural fibres	Advantages	Disadvantages
Cotton	High wet tear resistance High moisture absorption Biodegradable Good dyability	Poor temperature stability High combustibility Low chemical resistance
Wool	Good heat storage capacity Good dyability	Tends to felt Poor temperature stability Low chemical resistance Low scrub resistance
Jute, hemp	Low density	Average temperature stability
Flax	Low abrasiveness Good water absorption capacity High strength at low elongation Biodegradable Good dyability	Low chemical resistance

Table 6.4 Properties of textile sheets

Fibre differences	Brilliance	Bending stiffness	Packing density	Appearance	Feel	Clothing comfort
Refinement	+	+ + +	+ +	+	+ +	+ + +
Profile in cross section	+ + +	+	+ +	+ + +	+ +	+
Shrinkage	−	−	+ + +	+	+ + +	+ +
Crimp	+ +	+	+ + +	+	+ +	+ +
Moisture absorption capacity	−	+	+ +	−	+	+ +

In addition to the starting raw material, also geometric and physical properties determine the profile of the fibre. The left column of Table 6.4 lists the general geometric traits that differentiate fibres. The physical properties of the surface that is derived result from the composition of the fibres.

6.1.1.2 Production Process

Firstly, a yarn is manufactured from the spinnable fibre or the filament by way of different process steps. The following work operations must be accomplished:

- Blow room:

  - Opening the bales
  - Mixing the fibre flocks
  - Cleansing of the fibrous material

- Card room:
  - Break-up of the flocks
  - Parallelisation
  - Cleansing
  - Formation of a sliver

- Combing works:
  - Elimination of short fibres
  - Fibre parallelisation

- Stretching works:
  - Homogenisation
  - Arranging of the fibres
  - Mixing
  - Elimination of dust

- Spinning mill:
  - Refinement
  - Winding up

The yarn is processed further into a sheet. Three basic types are differentiated here:

Fabric:	Crossing of two or more sets of threads.
Knit fabric:	Mesh-like formation of loops of one or more sets of threads.
	• Knitted fabric, weft-knitted (crosswise yarns)
	• Warp-knitted fabrics (lengthwise yarns)
Non-woven material:	Sheet is formed by consistent, cohesive or adhesive joining of the fibres or threads.

Fabrics can be defined as textile sheets which are created by the perpendicular crossing of two or more sets of threads. The lengthwise system is called the warp threads and the crosswise system the weft threads. The way the two systems cross is termed the weave.

A fabric is characterised by a cohesive surface. Its strength-elongation characteristics depend on those of the threads used, as well as the density of the crossing points (Fig. 6.2).

Knit and interlaced fabrics are summarised by the term "knitted fabrics" Knitted fabrics are fabrics characterised by the formation of a textile sheet by mesh-like interlacing (the formation of loops) of one or more sets of threads. Knitted fabrics are characterised by an open surface and a high degree of deformation under the influence of minimal force. Moreover the sheets contain tensions that can lead to the edges rolling up.

Knit fabrics are created by a crosswise system of threads in which, individually and sequentially, loops are formed as the threads are interlaced or joined with a row of half loops (Fig. 6.3).

Interlaced fabric can be divided into weft knit goods and warp knit goods. Weft knit goods are a system of threads—similar to knit fabrics—that is worked horizontally

Fig. 6.2 Fabric structure

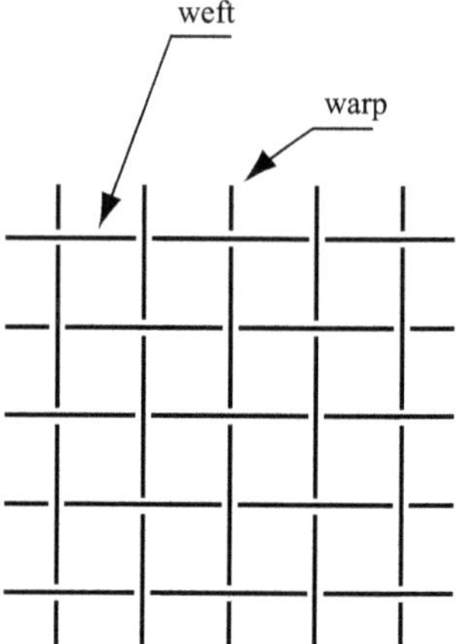

Fig. 6.3 Knit structure

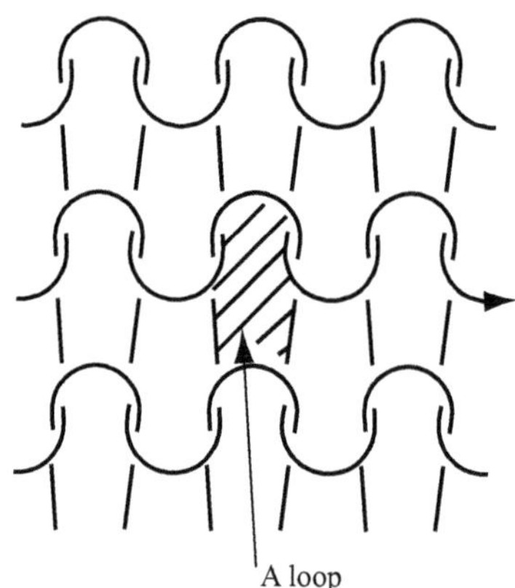

Fig. 6.4 Interlace structure

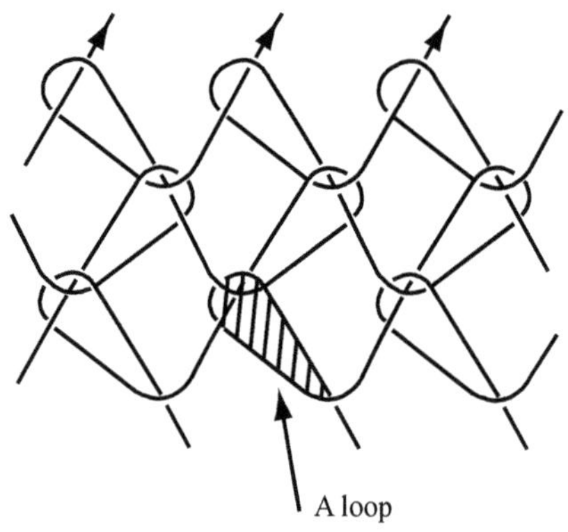

Table 6.5 Properties of textile sheets

Characteristics	Non-woven	Atlas fabric	Twill fabric	Multiaxial interlace	Knit
Tensile strength	−	+ +	+ −	+	−
Tensile rigidity	−	+	+ −	+	−
Drapability	+ +	+ −	+ −	+ −	+ +
Fibre volume content		+ +	+	+	+ −
Energy absorption capacity	−	+ −	+	+	+ +

course for course. In terms of the kind of binding, there is no difference between weft knit goods and knit fabrics, which is why both terms are often used synonymously (Fig. 6.4).

In contrast to knit fabrics interlaced fabrics are created by the simultaneous formation of multiple loops within a lengthwise system of threads. The connection between the parallel arranged threads is formed by a mutual, sideways loop.

The surfaces of non-woven fabrics range from smooth to voluminous. Their strength-elongation characteristics are very similar to those of the fibres or threads used for their manufacture.

The manufacture of non-woven fabrics is done in three production stages. The first stage is the formation of the fibres into the fabric. This can be done according to various principles which can lead to either a lengthwise or a lengthwise and crosswise orientation, or to a swirling of the fibres in the collection. The structure chosen for this fabric influences its characteristics in terms of its strength, for example.

In the second stage the fibres are strengthened into the non-woven fabric. To do this different methods are used such as needling with a needle machine or spraying with a binder.

In a third step the non-woven fabrics are finished. This can be performed by e.g. washing, dying or the application of chemical finishes.

The structure of the sheet has a strong influence on the stability of the substrate.

Moreover, in the case of one-sided coating, the textile is responsible for the characteristics of the underside of the product, in other words the wearing comfort. For example, roughened cotton in direct contact with the skin is more comfortable than unroughened.

The fabric structure forms a relatively cohesive surface, whereby the strength-elongation characteristics are virtually identical to those of the type of thread used. In the case of knitted fabric, the open surface and the high degree of deformation when subjected to a minimal force must be taken into consideration. That is why indirect coating methods are principally used in the case of knitted fabric. Use of tenter frames would deform the textile. The elongation capacity in the case of this kind of manufacturing is—depending on the formation of the loops—higher than that of the individual threads.

Non-woven fabrics have similar strength-elongation characteristics as the threads or fibres used. They are characterised by a smooth to voluminous surface. The manufacturing method used to make non-woven fabrics, the material thickness as well as the materials used provide information about the usability with respect to direct or indirect coating of these substrates (Table 6.5).

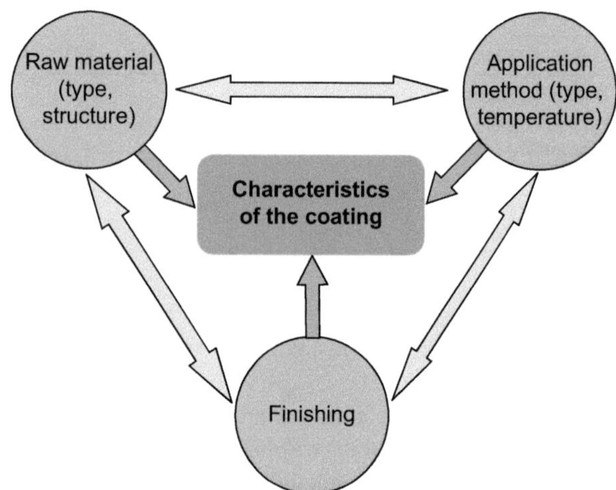

Fig. 6.5 Determination of the characteristics of the coating

6.1.2   Coating of the Surface

The coating has an influence on all surface characteristic s of the product. Among others, these include:

- UV resistance
- Weather resistance
- Heat resistance
- Chemical resistance
- Abrasion resistance
- Transparency
- Colour
- Dirt repellence
- Smell
- Antistatic properties

The intensity of these characteristics depends on the formulae; i.e. the chemicals chosen and their mixing ratios among one another.

6.1.3   Coated Textiles as Laminate Materials

Coated textiles can be classified as laminate materials. These materials are two-dimensional and comprise a composite made of two or more contiguous layers that, because of the intimacy their materials, can be viewed as single unit and can also be used as such (Fig. 6.5, Table 6.6).

Table 6.6 Advantages and disadvantages of different coating media

Coating media	Advantages	Disadvantages
PVC	Economical	Brittle (can be regulated by plasticiser additive)
	Slightly high-frequency	
	Fusible	Poor low-temperature flexibility
	High abrasion resistance	No chemical cleaning stability
	Good weather resistance	
PU	Wide range of variations in terms of feel and pliability	Limited weather resistance
		Not UV resistant
	Good buckling resistance	Not hydrolysis resistant
	High abrasion resistance	Expensive
	Microporous coating by coagulation method possible	
Acrylates	Good weathering and aging performance	Susceptible to solvents
		Limited buckling resistance
	Economical	Limited mechanical properties
Silicone	Good heat resistance	Expensive
	Good adhesive characteristics	

The combination of the coating and the textile carrier is responsible for the tear resistance, adhesive strength, combustion behaviour, and long-term durability of the end product. Thus, for example, in the case of the combination of PP yarn as carrier material and PVC as coating medium, despite its many attempts, the industry has been hitherto unsuccessful in achieving suitable adhesion. The bonding agents available on the market—as used in the case of chemical fibres—cannot satisfy demands in the case of the aforementioned combination (Fig. 6.6).

6.1.3.1 Composite Products Produced by Spread Coating

Carrier materials such as knit goods, fabric, paper, felt, non-woven fabrics and foils, as well as other materials in web form can be coated using the spread coating method. These can be processed into numerous kinds of different end products using pastes and coating materials made of polyurethanes (PU), polyvinyl chlorides (PVC), acrylates and silicones.

Products Made of PVC

- Handbags and suitcase materials
- Book binding materials
- Upholstery goods for the household and automotive industry
- Imitation leather for shoe and outer clothing
- Decoration materials
- Table cloths and bathroom curtains
- Tent and sail cloth materials
- Rain clothing and camouflage materials
- Truck and container tarpaulins

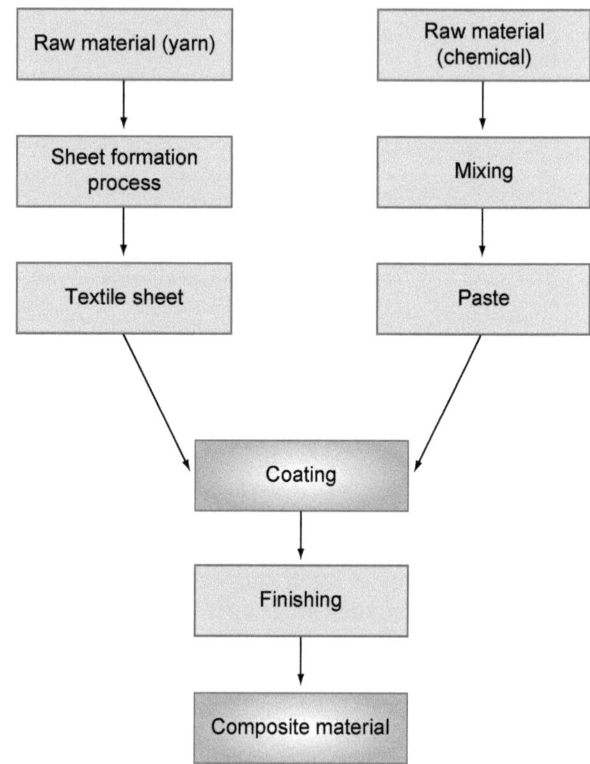

Fig. 6.6 The development of a laminate material

- Transport strap materials
- Vinyl wallpapers
- Vinyl floor coverings
- Sealing materials
- Technical articles

Products Made of PU

- Upholstery goods for the household and automotive industry
- Tent and sail cloth materials
- Imitation leather for shoe and outer clothing
- Decoration materials
- Rain clothing and camouflage materials
- Truck and container tarpaulins
- Sun protection
- Technical articles

Products Made of Acrylates

- Tent and sail cloth materials
- Rain clothing and camouflage materials

- Decoration materials
- Technical articles

Products Made of Silicones

- Tent and sail cloth materials
- Various kinds of stockings
- Airbags
- Heat protection clothing
- Technical articles

Note: Products for related areas of application which are created using various coating pastes differ among others in terms of softness, light-fastness, abrasion resistance, tear resistance, and compatibility with skin. As a result, the choice of a suitable coating and a carrier material is dependent on the specific requirements.

6.1.3.2   Knife Spreader Coating Machines for Manufacturing Imitation Leather

There are three kinds of imitation leather: imitation leather made of PVC, imitation leather made of PU, and semi-PU imitation leather which is made of both PU and PVC. Usually, all three kinds are coated using the reverse or transfer process. The only difference is that when using PU pastes, which are usually based on solvents, the coating system must have explosion protection systems.

Assuming a three-coat machine, during the production of imitation leather a grained carrier paper—coming from an unwinding station—is fed to the first knife coating machine. Here, the topcoat is applied, which has a thickness of approx. $20–185\,g/m^2$, depending on the intended use of the end product. In the subsequent drying tunnel, the web is pre-gelled at approx. $145–165\,°C$ (depending on the speed). The coated paper is subsequently cooled in the cooling unit, and then passed to the second knife coating machine where the foam coat is also applied using a spreading knife (10-mm chamfer). The thickness of the coat applied is approx. $220–1{,}000\,g/m^2$. Again, in the drying tunnel pre-gelling takes place with subsequent cooling in the cooling unit. In conclusion, on the third knife coating machine (doctor knife with 6-mm chamfer) the adhesive coat is applied with a thickness of approx. $65–185\,g/m^2$. Now, from the unwinder of the laminating unit the substrate—the actual carrier material—is now unwound and laminated in the still-wet paste. The chemical process (full gelling and foaming) is completed at a temperature of approx. $185–225\,°C$. After the web has cooled, the coated substrate is separated from the carrier paper in the delamination unit. The coated substrate and the original carrier material are each rolled up separately.

The setup shown in Fig. 6.7 is typical for high quality imitation leather. Besides this, there is also a thinner version of imitation leather which consists of only two coatings, and is used e.g. in bookbinderies. Here, the foam coat is omitted, which means that a two-coat machine would also be sufficient for this purpose (Fig. 6.8).

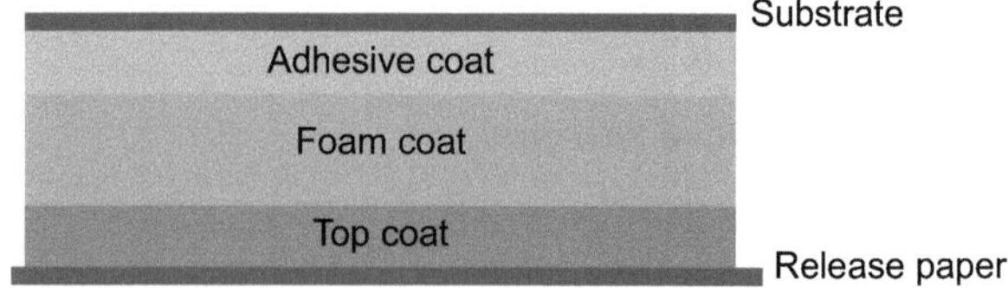

Fig. 6.7 Composition of imitation leather

Formerly, imitation leather was also manufactured using the direct coating method. Today, however, the industry produces almost exclusively using the transfer process, which offers several advantages compared to direct coating. Different kinds of embossed or also smooth release paper enable different surface structures and qualities without the necessity of subsequent embossing using a very expensive embossing roller. If skilfully handled, the release paper can be re-used up to 10 or, to some extent, even 15 times. Since the substrate is subsequently laminated and there are no direct tensile forces as a result, it is also possible to use elastic carrier materials such as knit goods. Imitation leather can also be manufactured this way without any carrier whatever (Brathuhn 1986).

Imitation leather is offered on the market at a cut width of approx. 1.40 m; more seldom at a width of 1.60 m. Consequently the common machine width is in a working range of 1.80 m, in order to compensate for the edge loss during coating and especially during laminating. The desired thicknesses are usually specified by the customers in the thicknesses 0.85, 1.0, 1.3, and 1.5 mm. In the case of bag-making, there is even sometimes a demand for materials with a thickness of 3.0 mm.

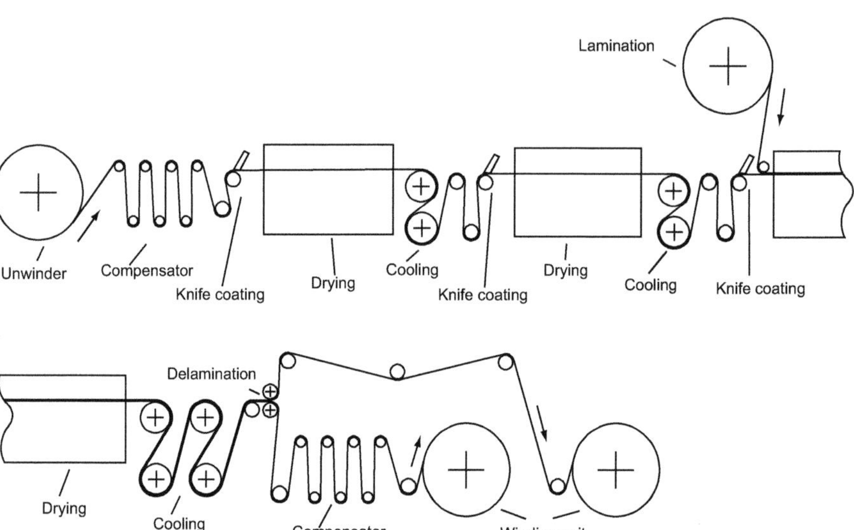

Fig. 6.8 Schematic illustration of an imitation leather production system

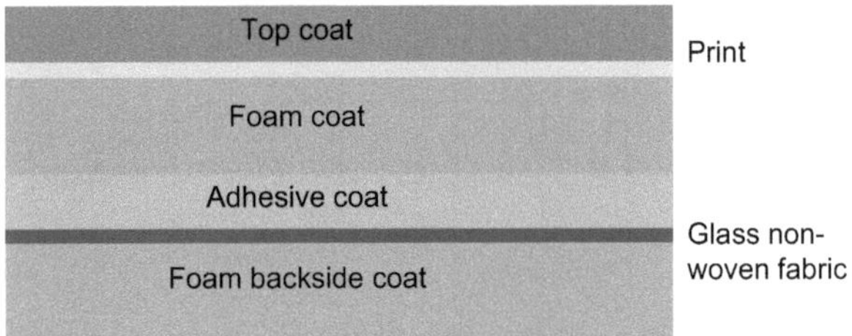

Fig. 6.9 Composition of a floor covering material

6.1.3.3 Knife Spreader Coating Machines for Manufacturing Floor Coverings

Floor coverings are always produced using the direct coating method. Usually two-coat or four-coat machines are used. Here, a four-coat machine is assumed because it provides higher output. A base carrier (glass non-woven fabric, board or a mixed substrate) is supplied by the unwinding unit to the first knife coating machine where the back coating (foam backside coating) is carried out with a coat thickness of approx. 80–225 g/m^2 (6 mm chamfer). The web is pre-gelled in the drying tunnel at 150–185 °C, subsequently embossed and cooled, and then fed by the turning unit to the second knife coating machine (Fig. 6.9).

On this knife coating machine, the adhesive coat is now applied using a doctor knife (10 mm chamfer) at a thickness of approx 200–650 g/m^2, and also pre-gelled in the tunnel at 150–185 °C, After cooling, in the third knife coating machine (6 mm chamfer) the foam coat is applied at a thickness of approx. 250–800 g/m^2. After drying and cooling the web is fed to the four-colour printer where printing is done. An inhibitor is mixed into the final printing paste, which prevents foaming of the foam coat in the areas that are crosslinked by it. In the fourth knife coating machine, the top coat (a transparent coat) with a thickness of approx. 80–250 g/m^2 is applied to the walking surface by a doctor knife (6 mm chamfer), and then gelled and foamed at approx. 185–225 °C in the drying tunnel. The finished web is wound up after cooling (Fig. 6.10).

The quality of a floor covering is determined primarily by the substrate used. The best quality can be achieved with glass non-woven fabric. In second place is special paper that is a combination of glass non-woven fabric and paper. In third place, pure board can be mentioned. Because the floor covering market varies drastically compared to other products in individual countries it will be covered in detail as follows.

Markets for Floor Coverings with Glass Non-Woven Fabric Glass non-woven fabric lends particularly good physical characteristics to floor coverings. Characteristics to be highlighted here are the outstanding strength, the recyclability, and the good adhesion of the glass non-woven fabric to the PVC. The different floor coverings are offered both with and without a back coating. Aside from the USA, where such qualities are not in demand, they are sold under the name of cushion vinyl. The thicknesses lie between 1.0 and 3.0 mm, and in special cases 5.0 mm.

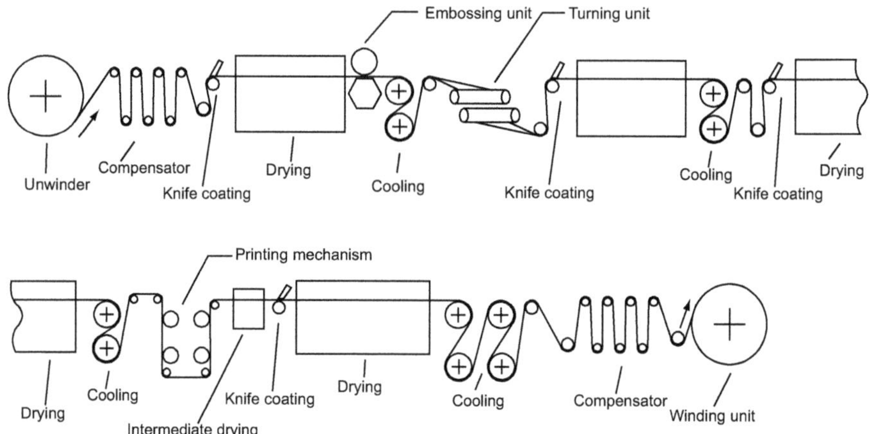

Fig. 6.10 Schematic illustration of a floor covering production system

Markets for Mixed Substrates (FÜR, FPS) This carrier material has characteristics similar to pure glass non-woven fabric, but its quality is somewhat poorer.

Europe: 1.2–2.0 mm thickness;   FPR 0.35 mm; 0.44 mm
Asia: 1.2–2.0 mm thickness;   FPR 0.35 mm; 0.44 mm

The material cannot be sold in the USA and CIS.

Markets for Floor Coverings with Board This floor covering has no back coating. The board used (Arjocast) is available in various thicknesses.

Europe: 1.0 mm;   thickness Arjocast 0.35 mm; 0.60 mm
Asia: 1.2–1.4 mm;   thickness Arjocast 0.50 mm
USA: 1.0–3.0 mm;   thickness Arjocast 0.50 mm; 0.60 mm
CIS: 2.0–3.0 mm;   thickness Arjocast 0.60 mm

Floor coverings are available in three widths: 2; 3 and 4 m, whereby the market shares lie at 50, 40 and 10 %.

6.1.3.4 Knife Spreader Coating Machines for Manufacturing Tarpaulins

The direct coating method is used to manufacture tarpaulins. The substrate, which is a polyester or polymide fabric, is fed to the first knife coating machine where, first, the base coat is applied to the underside. The coat thickness is about 20–185 g/m^2 (6 mm chamfer). Using tenter frames the substrate first goes through the drying tunnel, which works at temperatures of around 145–185 °C. The substrate spends some time in a calender in order to achieve a smoothed effect, before it is passed on to the cooling unit. Going through a turning station, where the substrate is turned 180°, it comes to the second knife coating machine where the adhesive coat is applied to the

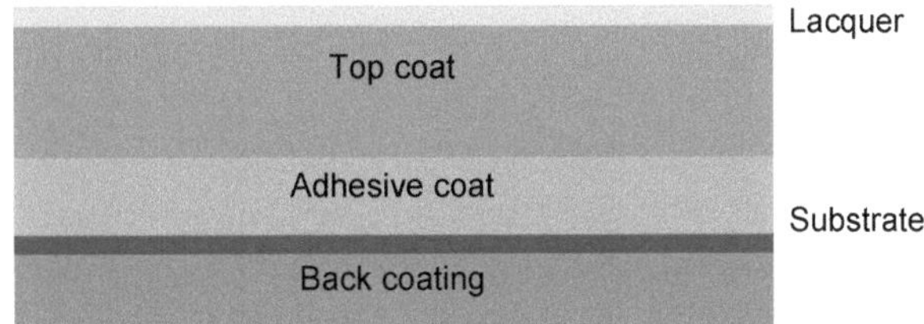

Fig. 6.11 Composition of a tarpaulin material

upper side. Drying and cooling are performed using the same method as in the first knife coating machine. The top coat is applied in the third knife coating machine, where full gelling at a temperature of approx. 185–225 °C has taken place in the subsequent drying and cooling processes. The final lacquering, as finish, is done by an engraved roller. Then drying and cooling are done and the finished end product is rolled up (Figs. 6.11 and 6.12).

Some tarpaulins exist which, in addition to the base coat of the underside, also have a top coat, or a top coat and a finish. But this occurs only in special cases.

The widths offered for sale in the case of tarpaulins lie in the range of 1.40–3.0 m. More than anything else the weight—with respect to area—is emphasised as a quality factor. The weight of high quality tarpaulins lies at 600–800 g/m^2.

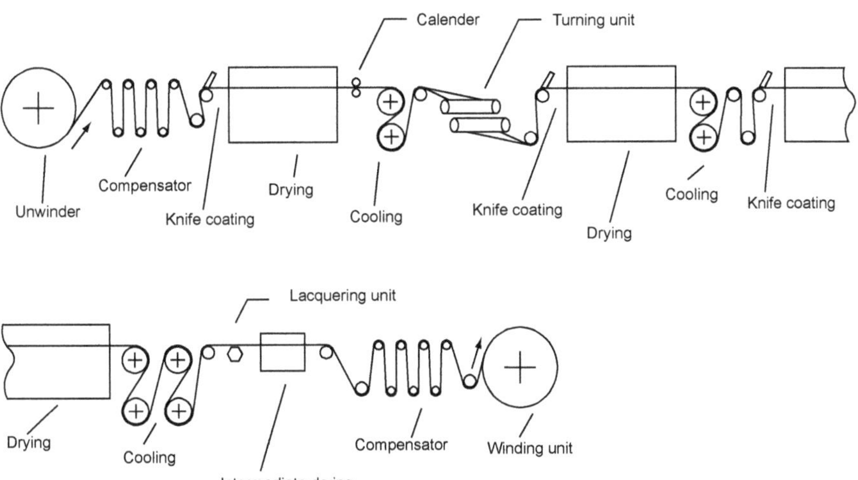

Fig. 6.12 Schematic illustration of a tarpaulin production system

Fig. 6.13 Composition of clothing materials

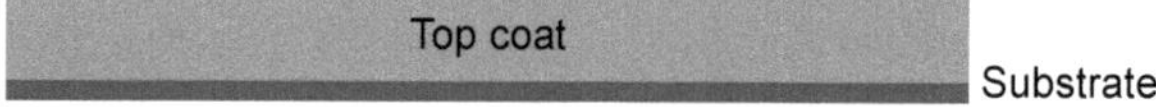

6.1.3.5 Knife Spreader Coating Machines for Manufacturing Clothing Materials

Clothing materials can be defined as light fabrics that are coated with a PU coating in order to achieve properties such as weather resistance and ease of washing.

A one-coat production system is required for this kind of end product. With the help of tenter frames, the top coat with a thickness of 10–$250\,\text{g/m}^2$ is applied in a knife coating machine, and gelled at 185–$220\,°\text{C}$. This is followed by the cooling and rolling up operations. The air knife system with a chamfer width of 6 mm has established itself as an application system.

For clothing fabrics the saleable widths of goods are 0.84, 1.50, 1.68 and 1.80 m (Figs. 6.13 and 6.14).

6.2 Medical and Hygiene Products

Surface functionalisation is a particularly important criterion in this area. Most products only come into being after they are coated with chemicals which lend the products a certain functionality. The individual products can be divided into the areas/markets described in the following section.

6.2.1 Product Description

Hygienic Products Here, one finds mostly low-cost products such as e.g. nappies, women's hygiene products, and mass-produced wound dressings. Quick absorp-

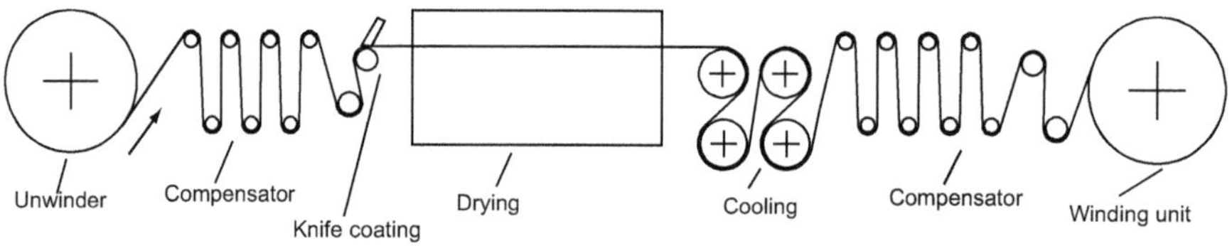

Fig. 6.14 Schematic illustration of a clothing material production system

tion and retention of fluids, and breathability properties are usually required in this instance.

Functional Textiles These products are often used in medical facilities. Examples are protective clothing such as operating room clothing, operating room coverings, but also hygienic consumables such as e.g. mattress underlays. The functions demanded here tend to be somewhat more manifold. Alcohol- or water-repellent, and super-absorbent are classic features.

Medical Applications In the case of these products, the coating not only finishes the carrier medium, it also applies the actual functionality. In the case of wound dressings or transdermal systems the necessary active agents are applied to the carrier medium with high precision. Examples of transdermal systems include nicotine or heat plasters, but also hormone plasters which are used for low-dosage contraception.

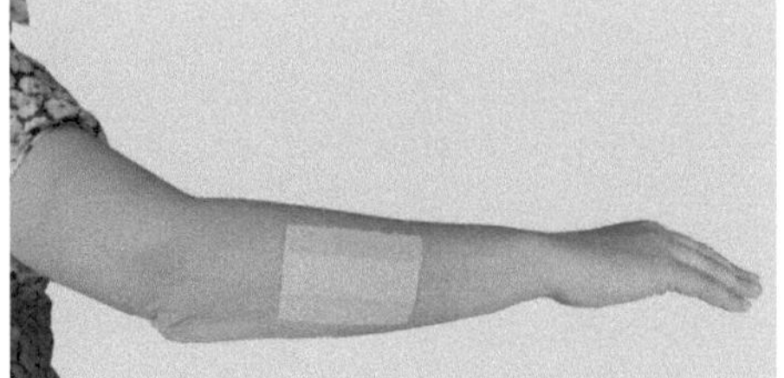

Other Products Since medical care prompts an endless range of different products, it is perfectly natural that many other coated products are available in this field. For example, for diagnosis purposes markers are produced by coating processes which indicate pathogens or reveal other indications of diseases in blood or urine. One new development is gelatine strips which are coated with certain active agents. These are simply sucked on or placed in the mouth on the gums where they can quickly and efficiently release their active agent. Even the lead foil that is used to provide protection against X-rays must be coated. In the same way individual filaments, yarns or threads are coated for medical use. Be it dental floss or surgical thread for sewing—these products also only achieve their necessary properties through coating.

6.2.2 Substrates and Chemistry

The substrates and chemicals to be used are determined by the desired effect. The following functions are those which one aims to achieve or strengthen by coating:

Medical applications	Strength elongation	Antistatic properties	Combustibility
Hydrophilic properties	Hydrophobic properties	Moisture transport	Oleophobic properties
Inorganic coatings	Soiling behaviour	Self-cleaning	Protective function

Usually only cellulose materials are considered for substrates in the area of hygiene since they have the best liquid absorption properties and their manufacture is one of the cheapest. Hydrophobic properties—or water-repellent surfaces—are achieved by using water or solvent based fluorocarbons. In order to achieve the opposite effect— i.e. hydrophilic properties—silicones, copolymers and so-called "wetting agents" are used to greatly increase the absorption capabilities of a substrate. Functional textiles are also usually intended as disposable articles so they are of lower quality. Cellulose materials and plain textiles are coated in order to be made alcohol repellent and antistatic, among others.

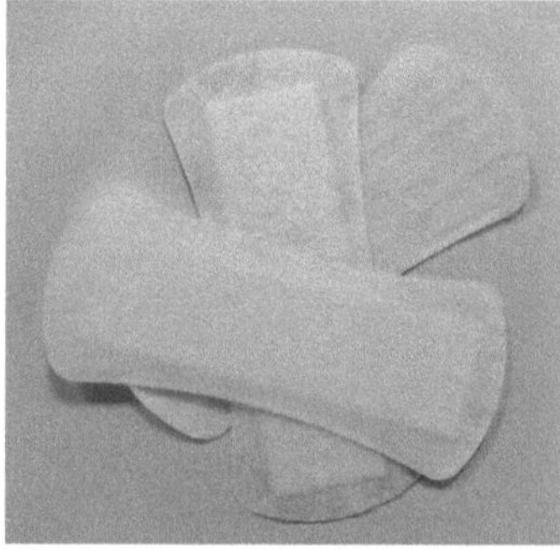

Other purposes are to kill bacteria, viruses, fungi, and algae. Chemicals prepared with silver nanoparticles are usually used for this. The medical applications predominantly include wound treatment and plasters with additional active agents. Various medical pastes are applied to different textiles such as cotton strips, gauze, and non-woven materials. Thus there are pain-killing, healing and preventative types of plasters such as ABC plasters and nicotine plasters, but also plasters with hormones that can be used as a form of contraception. The advantage of such so-called transdermal systems is that they can be stuck to the skin. There they release the medicine continuously and very precisely into the blood circulation system, without having to pass through the liver or kidney beforehand.

For other medical products almost any kind material in web form can be used: papers, etched foils, metal substrates, etc. Every form and size is conceivable—right down to a single thread.

6.2.3   Process Engineering

Individual products and product groups naturally require different processes. However, the following points can principally be named as typical for the production of all of these product groups:

- Application of thin layers with extreme precision
- Special surface treatment
- Purity in the coating process (GMP standard to some extent)
- Most meticulous of recording functions (Quality management),
- Flexible layout with various coating technologies
- Quality assurance systems
- Conformity with respect to all applicable legal aspects
  (e.g. environmental regulations such as ATEX, VOC, and similar).

Hygienic and medical products are usually manufactured at a work width of 1,000–2,200 m. In some cases, high-speed production lines with speeds up to 800 m/min. are used for low-cost hygienic products. In the case of high quality medical products, a production line speed of up to 5 m/min. can be rather on the high side.

Special attention is to be given to the purity of the product being manufactured because most of the products not only come into close contact with the body, but also cover open areas. That is why enclosed production plants, which cover the entire production process right to packaging of the individual product, are not unusual.

One potential production method shall be examined using the transdermal system (TDS):

There are two different types of such TDSs: a matrix TDS and a reservoir TDS. The matrix TDS is simply coated on the carrier like a film and can administer up to a maximum of 5 mg/day. In most matrix TDSs the active agent is contained in the plaster together with the adhesive. For larger quantities and for administering different drugs the reservoir TDS is used.

Principally all TDSs are packed between the carrier and the release paper. Some possible structures are shown here:

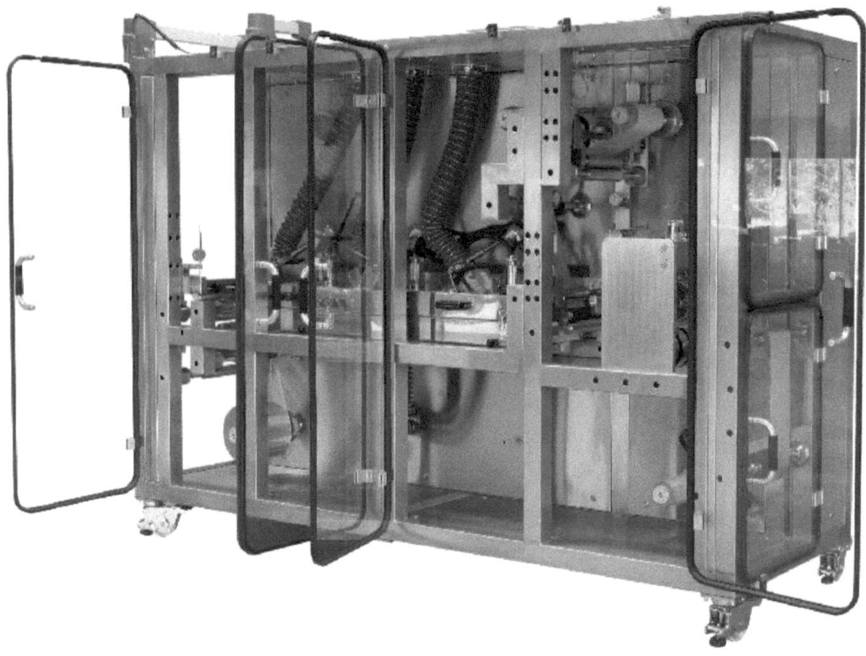

Fig. 6.15 Medical coating production system, GMP version

Matrix TDS	Reservoir (dry) TDS
Carrier	Carrier
Adhesive and medicine	Medicine
Release paper	Breathable membrane

As can already be seen in the illustrations, usually the carrier is the first to be coated with the active agent. After this, the adhesive is coated with the release paper. In a final step both sheets, carrier and active agent as well as release paper and adhesive are laminated together (Fig. 6.15). This means that the sequence in a typical coating system for medical products has the following structure:

1. Unwinder
2. Coating (see coating systems)
3. Calender or lamination
4. Drying (usually hot air and not IR or UV)
5. Cutting (edge trimming and cutting to size)
6. Winding up or packaging

Usually the following application systems are used:

1. Kiss coater, for a coating weight of 2–100 g/m^2, viscosity of 10–10,000 mPa
2. Slot die nozzle, for a coating weight of 2–300 g/m^2, viscosity of 100–18,000 mPa
3. Engraved roller, for a coating weight of 5–120 g/m^2, viscosity of 100–10,000 mPa
4. Knife systems, for a coating weight of 8–1,000 g/m^2, viscosity of 1,000–50,000 mPa

6.3 Prepreg

Prepregs are semi-finished products and are part of the processing chain in the manufacture of fibre-reinforced composite materials. A fibre-reinforced composite material is a material that is usually composed of two main components. On one hand it is a bed of a polymer matrix, and on the other it is a reinforcing fibrous material (Wikipedia—Fibre-reinforced composite 2008). Because of their low weight, fibre-reinforced composite materials are used as an alternative to metal constructions.

Depending on the length of the fibres, these fibre matrix materials are divided into the following products, which are produced in a thermoset or thermoplastic matrix.

Thermoset polymer matrix	Thermoplastic polymer matrix
SMC (sheet moulding compound): thermoset polymer matrix with fibre lengths between 25–50 mm	GMT (glass mat reinforced thermoplasts): composite made of thermoplasts with fibre mats
BMC (bulk moulding compound): polymer matrix with extremely short fibres (approx. 0.1–1 mm)	LFT (long fibre thermoplasts): thermoplasts with long fibres > 25 mm reinforced
Thermoset prepregs: polymer matrix with endless fibres, as rovings or fabric	Thermoplastic prepregs: higher quality thermoplasts (PA or PEEK) with endless fibres (Schürmann 2008)

Thermoset prepregs are explained in more detail as follows.

The strength achieved in the fibre-reinforced composite material is influenced by the length of the fibres. The longer the fibres, the better the properties. This way prepregs represent the cornerstone of quality and premium products in the manufacture of lightweight synthetic construction materials; in use they can withstand the highest stresses but have the lightest self-weight. They are thermoset semi-finished products with tailor-made properties for their respective use in the areas of automobiles and transport, wind energy, aerospace, ship construction, medical technology, construction, electronics, sport and recreational items.

6.3.1 Product Description

Thermoset prepregs—i.e. semi-finished products made of preimpregnated endless fibres—are made of a combination of carbon, glass or Kevlar fibres and epoxy resin, and are processed into fibre-reinforced plastics. Depending on the final area of use, the proportion of the weight of the resin in the final product is a maximum of 30–40 %. The prepreg matrix consists of a mixture of resin and curing agent, and in some cases an accelerator is also used (Wikipedia—Prepreg 2008). In addition to high temperatures, also pressure is required for curing. Usually this is done according to the temperature-pressure time progressions prescribed by the manufacturers. Curing can take place under atmospheric pressure (approx. 1 bar) but the quality improves with higher pressure. Autoclaves with up to 10 bar are used in the most demanding of situations. The high pressure can reduce the incidence of void areas, entrapped air, and pores which can generate stress peaks under load, and would thus have a negative influence on the properties and the interconnection of the end product (Schürmann 2008). However, low temperature prepregs are also manufactured, depending on the area of application.

Prepregs must be carefully stored in a cool environment, ideally one that is airtight and has a protective foil or covering. By refrigeration at temperatures as low as $-20\,°C$ premature reactions between the resin and the curing agent are prevented. The material must then be "thawed" before processing.

Special manufacturing processes and additives can be used to achieve a desired degree of tack.

The prepreg is positioned cold in a form and fully cured under pressure and a temperature of up to 170 °C. When needed, the required format can be achieved by placing several layers on top of one another. The tack allows several layers to be processed.

In the subsequent heating process the resin liquefies for a brief period and soaks the fibres completely again before it begins to cure (Lange Ritter 2008).

Prepregs are used as fibre-reinforced plastics. Their advantages are:

- High rigidity
- High tensile strength
- Low density
- Corrosion resistance
- High vibration resistance
- Low thermal expansion
- Low weight
- Extraordinary fatigue resistance
- Easy use
- Low maintenance costs wherever used (CG-Tec 2008).

A machinery concept for the manufacture of prepregs principally includes the application of epoxy resin or phenolic resin to the widest range of substrates.

6.3.2 Substrates and Chemistry

Various starting products are used for prepregs. These are mainly based on:

- Carbon fibres
- Glass fibres
- Kevlar fibres
- Other synthetic fibres

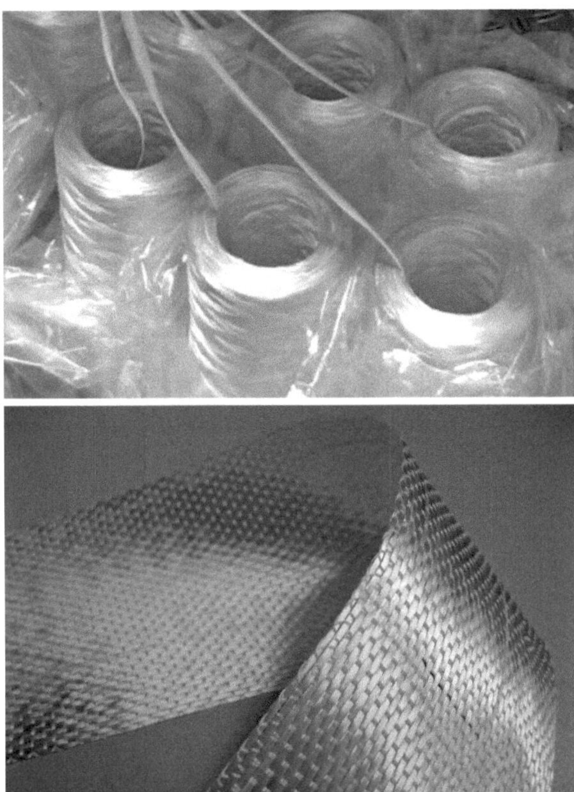

Fig. 6.16 Glass fibre bundles and fabrics made of fibreglass

Here, the substrates to be processed can take the widest range of formats, namely as:

- Fabric
- Uni, bi, and multidirectional fibre constructions
- Products that have many orientations according to the direction of tension required in the final use

However, the mentioned starting products are extremely sensitive in each of the formats. Before a coating process begins, the substrate must be placed in the optimum position by appropriate control and tension units (Fig. 6.16).

Here, the substrate must be fed to the coating process and the subsequent drying process, without having fibres shift out of the individual rovings and out of a fabric. The woven substrate's resilience to shifting is minimal, and causes a great deal of problems. The substrates must be very carefully and sensitively guided because there is virtually no physical cohesion between the fibre bundles.

Another important requirement on the processing of materials for prepreg products concerns the chemicals. The coating substance's influence defines the process and also the quality and performance of the end product. The mixture comprised

Fig. 6.17 Winder

of resin and curing agent forms the matrix. Various kinds of resins are used for coating:

- Solvent based or water based epoxy resins
- Materials in powder form,
- Hot melts
- Phenolic systems
- Polyester, vinyl ester and acrylic resin systems

6.3.3   *Process Engineering*

The way the substrate material is fed depends on its structure. Sheet substrates are fed to the coating process by way of unwinders, and rovings by way of appropriate creels (Fig. 6.17). The size of the creel is determined by the number of rovings to be coated, which are fed to the machine parallel to one another.

There are principally many different technologies to coat prepreg products; here a dipping and a coating process using a heated knife are the methods of preference.

Among the coating methods used to coat prepreg semi-finished products are:

- Coating by dipping (see Sect. 3.2.7)
- Coating using the knife technique (often heated) (see Sect. 3.2.1)
- Coating using roller systems (see Sect. 3.2.2)
- Coating by slot die application (see Sect. 3.2.6)
- Coating using powder (see Sect. 3.2.3)
- Coating using hot-melt (see Sect. 3.2.4)

The selection and an assessment of the choice of an appropriate coating method must be done according to the chemicals, the application weight required, and the

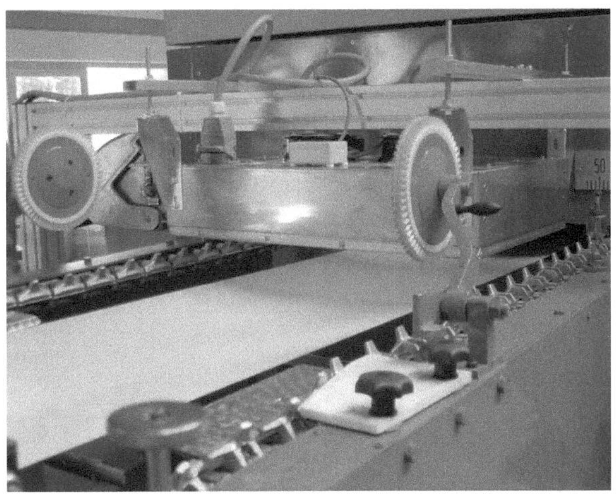

Fig. 6.18 Infrared drying and hot air

viscosity of the coating material. Also decisive for the decision is the choice between full impregnation of the material or merely a one-sided coating.

Problems arise in the processing of the starting products for the following reasons, among others:

- Broken fibres
- Insufficient soaking and covering so that not all fibres are covered, regardless whether rovings or a cohesive sheet is being coated
- Critical vapours such as formaldehyde or phenol components during the drying process

Besides the coating, drying is also a very important component of a coating system. Generally, in order to achieve high quality end products, perfect temperature settings and control are important for the drying and curing process. Methods used are:

- Drying using hot air (Sect. 3.5.1.1)
- Drying using infrared rays (Sect. 3.5.1.2).

Depending on the coating raw materials and the coat thickness, both methods are used individually or also in combination. Convection dryers are used when processing solvent-based raw materials and when consistent air flow is required. Infrared technology can accelerate the drying process because the rays affect the interior of the substrate, and the heat distributes quickly. It represents the ideal drying method for high production speeds and in the case of water based coatings (Figs. 6.18 and 6.19).

The following important parameters must be taken into consideration for prepreg manufacturing when constructing a specified system layout.

- Individual winding technology
- Sensitive tension control in all process steps

Fig. 6.19 An installed coating system for prepreg

- Adapted coating methods, dependent on the application weight of the coating substance, the viscosity, and the feeding method
- Drying technology, dependent on the coating type (solvent or water based)
- Optimum air extraction system

6.4   Thin Film Coating (In the Field of Electronics)

Process Steps As introduction, first a schematic illustration of a simplified process chain will be given, as can be assumed as good example of a thin film coating system.

Since it is very possible that the process engineering can vary, even in a seemingly technologically well defined field such as thin film coating, the process shown in Fig. 6.20 makes no claim to completeness, but rather should serve merely as guide. The individual elements of the process chain will be successively shown and explained as follows.

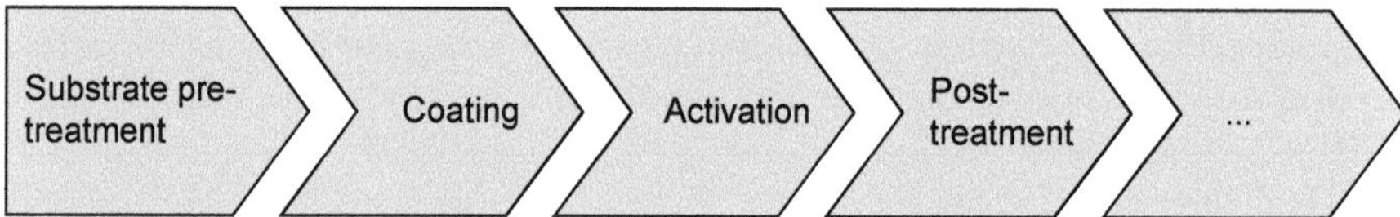

Fig. 6.20 Process chain for thin film coating (simplified diagram)

6.4.1   Substrate Pre-Treatment

Pre-conditioning of a substrate can be performed particularly by influencing the surface energy. Two widely established methods are briefly described below. Another way to alter substrate properties before the actual coating process, or respectively, to optimise the substrate surface with respect to the coating process, exists in the form of different methods of cleaning. However, because of their subordinate relevance to this chapter, they will not be explained at this point.

6.4.1.1   Corona Treatment

The so-called corona treatment (also called air plasma treatment) is a surface treatment process that is aimed towards the improvement of adhesion characteristics of substrates. This way, by modifying surface energies of substrates, such as e.g. paper or polymer-based films and foils, as well as other carrier materials, by lifting the so-called dyne levels, adhesive characteristics and surface crosslinking properties can be lastingly influenced.

Generally, corona systems for surface treatment can be used with virtually all kinds of substrates, including 3D objects. The corona treatment has established itself as industry standard when dealing with the modification of polymer films, and also plays an important role in the extrusion and converting industry.

6.4.1.2   Atmospheric Plasma

The atmospheric plasma method is principally very similar to the corona treatment; however a few differences between the systems exist. Both systems use one or more high voltage electrodes which positively charge ions in the surrounding air. Nonetheless, in the case of atmospheric plasma, the ratio of the oxygen molecules adhering to the molecule ends of the respective substrate is up to a 100 times larger. Resulting from this increased oxygen content is a much greater bombardment by ions. This results in stronger adhesive characteristics in the material, and increased absorption properties for liquid coating media (Wikipedia (1) 2009).

Practical experience shows that, over time, the effects of the above-named pre-conditioning methods are subject to a reduction which is independent of the substrate. This means that the longer the time between the pre-treatment step and the coating process, the less effective the treatment in terms of an improvement in adhesive characteristics. This is taken into account in the construction of coating systems by integrating the corresponding components in *one* system (inline) in order to be able to carry out the coating process directly after the pre-conditioning process.

6.4.2 Coating

6.4.2.1 Coating Precision/Requirement Profile

Coating precision is given the highest priority especially in thin film applications since here, because of the thin overall coat thickness, even the smallest deviation could have a considerable influence on the functionality of the component or end product. In other words, the more precise the coating is applied in lengthwise and crosswise directions on the web, the higher the quality of the end product. In many cases, the desired function cannot even be achieved unless a defined coat precision is guaranteed. Thus, Edward D. Cohen points out in his work "Modern Coating and Drying Technology": "At best, we can control to about $\pm 2\%$ of the target coverage, using gravure, precision slot, slide or curtain coatings. Reverse roll coating with precision rolls and air bearings, can do as well. Other coating methods tend to give poor control" (Cohen 1992). At the latest, through Cohen's findings, it becomes clear that only a small number of application systems are at all capable of satisfying the extreme demands of a highly precise and homogeneous coating in the lower micrometer range (of the wet film). Without deviating far from Cohen's recommendations, this chapter will continue to shed light on the above named systems. "Reverse-roll" and "slide" and "curtain coating", however, will not be considered in this context.

6.4.2.2 Typical Application Systems

The main intention of this chapter is to highlight application systems which are typically capable of applying extremely thin and simultaneously homogeneous coats/films. Included here in particular are coating systems such as substrate-independent (because they do not come in contact) slot die technology, the so-called kiss coater system, and coating systems based on engraved rollers. For a deeper look at the systems mentioned, the study in the third chapter of Cohen is recommended (Cohen 1992, p. 25, "Basic elements of coating systems" [German: "Grundelemente von Beschichtungsanlagen"]). Here, detailed descriptions each of slot die and pouring technologies (Sect. 3.2.6), kiss coater systems (Sect. 3.2.2.4) and gravure roller systems are included (here: engraved rollers, Sect. 3.2.2.3). In addition, a brief characterisation of the three most important coating systems for applying thin coats follows below:

I. The Slot Die System Generally, nozzle or pouring technology, under which slot die systems can be subsumed as a type, provide a way of applying liquid coating media without making contact, so they are substrate-independent. A nozzle system typically consists of an application head (here the slot die), the feed unit in the form of pump technology with reservoir for the purpose of conveying the liquid medium, as well as an appropriate hose connection. In this way, the technology provides a self-contained system, which can be an advantage especially in the case of solvent-based coating media because premature evaporation (vaporisation) is prevented, and

Fig. 6.21 Slot die coating system

exogenous effects (impurities) cannot have an influence before the application takes place. A further advantage of this technology lies in the possibility of achieving striped or intermittent coatings by making appropriate modifications to the nozzle unit. The coat thickness is controlled by the mass flow being fed, the speed of the web, as well as the size of the space between nozzle unit and substrate, and the design of the nozzle head. A precisely constructed slot die system is characterised by its excellent precision when applying thin coats so that, when used in combination with low-pulse pump technology (e.g. a gear pump), it is predestined for the application of thin coats (Fig. 6.21).

II. The Kiss Coater System The so-called kiss coater belongs to the family of roller application systems and is also useful for precise, thin coat applications. At this point differentiation will not be made between kiss coater and so-called micro-roller systems, because the systems are very similar and the terms are often used synonymously.

The principle of the kiss coater provides for application of the coat by an application roller, which transfers the medium to the web, and by choosing a correspondingly small, virtually tangential angle of contact, which just skims the web lightly (hence the term "kiss"). Here, the application roller draws the coating medium from a reservoir with small dimensions which is situated underneath the roller. The guide rollers, by way of which the proper angle is set, are vertically adjustable. The system also provides a high degree of flexibility because by varying the contact angle of application roller, very thin coats as well as thicker coats with a higher range of viscosity can be achieved (Fig. 6.22).

III. Engraved Roller Systems As with kiss coaters, these systems belong to the family of roller application systems. This means that here, too, the application of the coat is achieved by direct contact of the application roller with the web (the substrate). The difference compared to other roller application systems consists mainly of the ability to define the coat by a pattern or engraving on the application roller (Fig. 6.23).

Fig. 6.22 A micro-roller
coating system

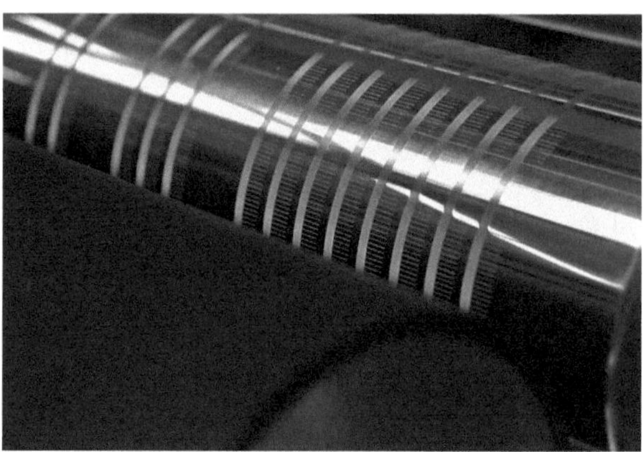

Fig. 6.23 Engraved roller
system

The pattern can be optimized with respect to the medium worked with by using advanced production technology such as micro engraving, or similar.

This also makes it possible, using a principally simple application system design, to achieve thin coatings. Just as in the case of the kiss coater, quality and precision when using an engraved roller system depend to a high degree on the production precision of the rollers and the precision of the mountings.

6.4.3   Activation

In the context of the conditions and assumptions laid out in this chapter, liquid application is typically followed by the activation of the chemical components or establishment of the integrity of the coating medium in whatever form this may take. Since, in most cases, the necessary thing to do is to uniformly evaporate the dissolvers (e.g. water or alcohols such as methanol or ethanol) via the introduction of thermal energy, the means available here is often drying in an oven, based on the principle of convection (Cohen 1992, p. 69 et seq.). Moreover, in typical coating media in the field of thin coatings, often there are UV reactive components which only cure fully when they are subject to UV light. There are also combined processes which are not unusual such as evaporation of dissolvers by convection in the first step and subsequent curing by UV radiation in a second, downstream step, and these are also taken into consideration in the construction of production systems. Related information on the background of UV technology can be found in Sect. 3.5.

The possibility of drying using infrared (IR) radiation will not be discussed at this point. Because of substrate composition and the frequent presence of alcoholic dissolvers in typical media of thin coatings in the area of electronics, and the problems associated with using an IR source, this method of drying is less relevant.

The two systems first mentioned (convection and UV) are briefly explained as follows, with a focus on their importance for the area of thin film coating.

6.4.3.1   Thermal

An exhaustive detailing of possibilities offered by thermal drying methods can be found in Sect. 3.5, starting on page 69. As follows, aspects which are of particular importance for the drying of thin coats are pointed out for the sake of example. Supposing that drying is carried out as follow-up to an application process, one assumes a precisely applied wet film of a specific coat thickness (usually in the one or two digit micrometer range) for which the intention is to dry it uniformly without negatively affecting the wet coat. Depending on the composition of this coat, excessive air speed/air volume, for example, can swirl or "blow away" the coat, which would lead to an inhomogeneous overall coat, and be unusable. Also the temperature range in which drying is to take place must be defined as precisely as possible. The temperature tolerances of the carrier materials (often polymers) as well as the requirements of the coating medium must be taken into consideration. It is not unusual that setting up a system of distribution in different temperature zones is recommendable, in order to achieve optimum drying conditions by way of a temperature curve generated by separately controllable quantities of air from above and below.

6.4.3.2   Reactive (UV)

Formulated compactly in a way appropriate to this chapter, reactive activation means the drying or the chemical curing of a coating medium by way of ultraviolet radiation. This process occurs through the photo-chemical reaction triggered by UV radiation, whereby a solid film is generated by crosslinking or polymerization. An essential ingredient of a coating medium suitable for UV activation is the photosensitive additive: the so-called photoinitiator. In addition to the high production speeds that can be achieved, UV curing is characterised by comparatively low heat radiation on the substrate (particularly interesting in the case of polymer-based foils), good surface quality, and that it immediately allows further finishing.

6.4.4 Substrate Post-Treatment (and Subsequent Processes)

After a possible pre-conditioning process has been applied to the substrate, and after the coating itself and the thermal and/or UV reactive activation (drying/curing), often additional processes follow. These can be manifold and are usually carried out to finalise the semi-finished product or end product. Typical examples here are a cooling or relaxing step whose purpose is to cool the substrate to a specific temperature before winding up, or also subsequent cleaning or etching processes. For example, subsequent processes can be the die cutting of specific geometries, side or centre cuts, the lamination of one or more substrates, or the application of a protective foil before winding.

Typical Substrates and Coating Substances Because of the high degree of coat precision which is typically demanded in the application of thin, wet coats, also the range of substrates to choose from is correspondingly limited. This means that the coat carrier (the substrate) must be a carrier with only minimal thickness tolerance levels in order for it to make sense to apply such a thin coating. Edward D. Cohen points out this requirement in his work, already in his introduction to drying thin coats: "The discussion will concentrate on those coatings that retain their integrity, have an impermeable substrate, and do not penetrate the web" (Cohen 1992, p. 267). For this reason, the only substrates relevant at this point are carrier materials conceivable as being suitable for manufacturing flexible electronics. To be precise, coating on polymer carrier foils or metal foils is assumed, and a quality in the manufacture of these substrates is assumed (low fluctuations in thickness in lengthwise and crosswise dimensions) which satisfies the requirements of highly precise coating methods.

The choice of the coating media (liquid) suitable for thin film coating is mainly determined by one decisive factor: the viscosity. Related information on rheologic properties can be read in Cohen in Sect. 3.2 (Cohen 1992, "Rheologic Fundamentals"; German title "Rheologische Grundlagen"). It is generally applicable that media in the low to medium viscosity range are suitable for thin film coating, whereby a series of other influencing factors must also be taken into consideration, such as e.g. the solids content or the type and number of solvents contained. An optimal coating system can be determined only after all physical and rheologic factors have been clarified. However, in practice, even if corresponding information is available, preference is given to empirical verification by appropriate laboratory tests, in order to achieve the best coating results with the means at hand. It should be briefly noted that changes to shear forces as influencing factors on the coating results have been consciously excluded here.

Concerning the type of coating substances, often layers are created that have conductive properties which are intended for use in end products such as e.g. touch screens. A further example is boundary/barrier layers such as those used in flexible photovoltaics. Most thin film applications applied as wet chemicals have the following in common, namely that they are intended as economically sensible substitutes for an established, but elaborate and expensive production process, usually found in the area of vacuum deposition technology. Research in this area is correspondingly active worldwide and well established.

Nanoscale function layers (i.e. coats produced by wet chemicals) represent a small peculiarity; these can form layer thicknesses in the upper to medium nanometre range after activation. Such extremely thin coats cannot be achieved by conventional application systems that were conceived for roller-to-roller coating of flexible substrates. Nevertheless there are ways, using properly prepared wet chemicals (coating media), to make advances into these spheres. One of these is the application of so-called sol-gels, whose production process is briefly described as follows:

> The sol-gel process is a method to manufacture non-metallic, inorganic or hybrid polymer materials out of colloid dispersions, so-called sols. The starting materials are also known as precursors. Out of these, fine particles are created in initial basic reactions in a solution. Special further processing of the sols makes it possible to create powder, fibre, membranes or aerogels. Because of the small size of the initially created sol particles in the nanometre range, the sol-gel process can be viewed as a part of chemical nanotechnology. (Wikipedia (2) 2009)

This way, sol-gel particles which have been enriched with solvents can be applied in the micrometer range using the application systems previously mentioned in this chapter, and after the activation process (usually after drying), form nanoscale functional layers.

Chapter 7
Characteristics and Applications of Plastisols and Additives

7.1 Introduction

Besides the familiar PVC suspension resins with a large palette of different products ranging from rigid profiles and tubes to soft garden hoses or electric cable coatings, another category of resins, so-called dispersion resins, exist, which are processed using completely different technologies.

While suspension resins made of powder or pellets are processed at high temperatures, these dispersion resins are changed at room temperature as liquid compound phase, the plastisol. The plastisol is made of a mixture comprised of a liquid component, a plasticizer and a PVC dispersion resin. Consequently, the finished products are always bendable in nature. These dispersion resins comprise an estimated 10 % of overall PVC manufacturing.

7.1.1 History

The history of plastisols began in Germany when the company IG Farben focused on this area. Commercial production began in the year 1938. During the Second World War Germany produced 3,000 t of plastisol annually. The first use was for the substitution of a dissolvable coating in coated fabrics. Later, soles for shoes and foils were produced.

7.1.2 PVC Resins for Plastisol

PVC resins are typically manufactured by emulsion and microsuspension polymerisation. Compared to suspension resins, their particle size is considerably smaller, which results in a non-free-flowing powder. On the exterior, a suspension resin resembles crystalline sugar; paste resins, in contrast, resemble flour. Most resins are

A. Giessmann, *Coating Substrates and Textiles,*
DOI 10.1007/978-3-642-29160-9_7, © Springer-Verlag Berlin Heidelberg 2012

homopolymer; others, in contrast are copolymer. Generally VC-VAC are produced. Important characteristics of the paste resins are:

- Their particle size and particle size distribution, primary and secondary
- Their type of the emulsifier used
- Their plasticiser absorption
- Their K-value

The primary particle size, which is kept on the latex level during polymerisation, determines the viscosity.

The secondary particle size, which is determined by the drying method and the milling process, also influences the viscosity and beyond that, also the aging of the paste. In contrast, the type of emulsifier influences the viscosity, the foaming capacity, and the water absorption properties, etc.

The amount of plasticiser absorbed is in fact the amount of plasticiser that is necessary to create a workable viscosity of the paste. The *K-value*, the intrinsic viscosity, and the viscosity index, which primarily represent the molecular weight, all have an influence on its ability to foam, and especially the mechanical properties of the end product.

Most resins are polymerised in autoclaves. In this autoclave the latex, which is composed of an emulsion made of P of VC primary particles in water, is formed in the most stable concentration possible. In order to create the PVC powder, the water must be removed from the latex by evaporation. This process is performed in large dryers. The latex is pulverised at the top of the dryer, and using a flow of hot air it is centrifuged downwards. The water of the latex drops vaporises and leaves a cenosphere. The cenosphere is a hollow, spherical shell consisting of sintered primary particles of the latex. After that the powder, consisting of cenospheres, is milled. A definitive PVC powder is created, made of multiple cenospheres, cenosphere particles, and primary particles in a secondary particle size. In a mixture with the plasticiser, the cenospheres and the cenosphere particles, depending on the degree of sintering and depending on the temperature of the dryer air, are partially or fully desegregated.

7.1.3 What Is a Plastisol?

We already touched on this area when we discussed a liquid and so-called plasticiser absorption. The recognised definition is: A plastisol is a *dispersion* of PVC resins in a plasticiser. That is why the term dispersion resin is used; in comparison the term sol in the word plastisol makes reference to the chemistry of colloids, the colloid state of a system in two phases: a continuous phase (the plasticiser) and a dispersed phase, the PVC. In fact, the simplest plastisol is an E-PVC, which is finely distributed in a plasticiser with a necessary quantity of a thermal stabiliser. Starting with a solid it is necessary to mix in a liquid—the plasticiser—to create a paste with low to high viscosity. We will see later how it is possible to give an end product different characteristics by the way the plastisol is formulated. After a paste has been changed via into its final form via a variety of technologies, it is cured and melted at the right temperature. The result after cooling is the desired, homogeneous end product.

7.2 Rheology

When we speak of liquids with different consistencies ranging to a paste, one must also naturally speak of rheology. Rheology is the branch of science that deals with deformation and the flow of matter under load. The way a plastisol flows—i.e. its rheologic characteristics—influences its behaviour in practical applications. Depending on the application (coating, dip lacquering, etc.), a certain rheology of the plastisol is required.

Let us discuss some theory.

7.2.1 Viscosity According to Newton

Let us assume a plane β, and at a distance h to this a plane section with a surface S parallel to β. Imagine that β is fixed in place, and that we exercise a force F parallel to β. Under the influence of the force F the section S is to move parallel to β at a velocity of V. Newton established the following principle:

- The fluid layers between the plane β and the S-section are to be moved with velocities between 0 (the velocity of the plane β) and V (the velocity of the section S). Hence $v = Vx/h$ applies, whereby x is the distance of the observed fluid layer from the plane β.
- A proportional interpretation exists between the shear stress F/S, which is influencing the fluid, and the shear rate v/x.

Thus follows $F/S = \eta v/x$ or $dF/dS = \eta dv/dx$, whereby τ, by definition, is the viscosity of the fluid and is equal to the shear stress divided by the shear rate. This viscosity is a material property which, according to Newton, is a constant. Shear stress is generally symbolised with the letter τ (tau) and shear rate, D.

Consequently, the following applies:

$$\tau = \frac{F}{S} = \frac{dF}{dS} \quad \text{and} \quad D = \frac{v}{x} = \frac{dv}{dx}$$

and according to Newton

$$\tau = \eta D \quad \text{and} \quad \eta = \frac{\tau}{D}$$

7.2.2   Shear Stress $\tau = F/S$ is in dyn/cm

For shear rate D, the following applies: $D = v/x$ in cm/s/cm, thus s^{-1}.

Viscosity $\eta = \tau/D$ in dyn s/cm is called the poise (P). The actual dimension in Pa s is 10 P.

By definition, the consistency of a material follows Newtonian law. Consistency is measured using its viscosity. Substances that obey Newtonian law are considered more or less constant, depending on whether their viscosity is more or less high. After only one century—i.e. at the middle/end of the eighteenth century—it was possible to verify that certain substances follow Newtonian law. In tests in liquid flow through capillaries, Poiseuille established a law where liquid flow through a capillary is determined as function of the capillary size and the pressure difference between the two capillary ends. Poiseuille's law assumes that the viscosity of a fluid obeys Newtonian law at a constant η-value, corresponding to τ/D. Most pure fluids with a low molecular weight (water, alcohols, light oils. . .) obey Newtonian law, but some fluids do not. In the case of the latter we have no proportionality between τ and D. Here, the viscosity does not obey Newtonian law. These fluids are not called Newtonian fluids.

7.2.3   Non-Newtonian Fluids

Let us describe the rheologic behaviour of non-Newtonian fluids by imagining a curve that demonstrates the relative variations of τ and D: $\tau = f (D)$ or $D = f(\tau)$.

The consistency of non-Newtonian fluids at every shear rate is measured using the corresponding viscosity coefficients; this depends on the shear rate. Three types of non-Newtonian behaviour can be described:

a) Plastic Bingham termed plastic as a material that does not flow until the shear stress, tau, has reached the value τ (c), i.e. the lower yield point. But once the lower yield point has been exceeded, the plastic flows in such a way that an increase in shear rate D is a linear function of the shear stress τ.

b) Pseudo-Plastic If the curve $D = F (\tau)$ is not a straight line, but rather a curve with a convex bend towards the axis τ, then the substance is called a pseudo-plastic, with or without a lower yield point. In this case the viscosity coefficient drops with increasing shear rate. It is now no longer possible to describe this behaviour using one parameter or two parameters as in the case of Newtonian substances or Bingham plastics. It is necessary to draw the curves.

c) Shear-Thickening Substances If the curve $D = f (\tau)$ demonstrates a convex bend towards the D axis, then the substance is described as shear-thickening. The viscosity coefficient increases with D.

7.2.4 *Rheologic Curves*

The curves $D = f(\tau)$ and $\eta = f(D)$ are mutually complementary. For our description we will use the curve $\eta = f(D)$ because it clearly improves the pseudo-plastic behaviour and shear-thickening behaviour, which are of great importance for coating.

7.2.5 *A Look at Actual Practice*

We have observed how plastics demonstrate different viscosities under different conditions, and we have also looked at shear rates. In the case of the shear of a material, different layers of the material move at different speeds.

Now imagine a stack of cards. The relative movement speed of the layers is only one factor of the shear rate. Another is the distance between the layers affected by shear. When you spread butter on bread using a knife the shear rate is dependent on the speed of movement of your knife, and on the distance between the blade and the bread. If you now imagine you are working in a coating plant and you are using a knife spreader to spread plastisol onto a carrier paper at a speed of 12 m/min. and a distance of 0.08 mm between the paper and the knife, then shear rate is expressed by a simple equation:

shear rate (in s^{-1}) = speed of the paper (in mm/s)/distance between knife and paper (in mm).

Consequently, the following applies:

$$\text{Shear rate} = D = \frac{200 \text{ mm/s}}{0.08 \text{ mm}} = 2{,}500 \text{ s}^{-1}$$

Accordingly, the following applies: with increasing coating speed, the shear rate also increases.

With increasing coat thickness (distance between knife and paper) the shear rate decreases.

Naturally, this also applies the other way round.

The concept of viscosity is known and generally accepted. Viscosity can be described as follows:

$$\eta(\text{viscosity}) = \frac{\tau(\text{shear stress})}{D(\text{shear rate})}$$

As you know, τ is the resistance to the acting force (or to put it another way, the force that has caused the shear).

Since it has been indicated that some liquids can demonstrate different viscosities at different shear rates, a diagram that illustrates viscosity compared to shear rate demonstrates the rheology of a plastisol.

The following can be said:

- Newtonian behaviour: Viscosity remains the same in all shear rates.
- Pseudo-plastic: Viscosity drops with rising shear rate.
- Shear thickening: Viscosity increases with shear rate.

When you look at your viscosity curve, you should be able to recognise the viscosity of your paste in the context of your working conditions. Later we will touch on the importance of this material.

7.2.6   Thixotropy and Curing

Two further important rheologic behaviours of a plastisol are thixotropy and curing.

7.2.6.1   Thixotropy

Thixotropy is the composition of a substance whose viscosity decreases over time because of a constant load. In the case of a constant shear rate, the viscosity drops over time and tends towards a threshold value. This effect is reversible. If the viscosity of a thixotropic substance is determined first with increasing shear rate, and then with decreasing shear rate, then the curve of the decreasing shear rate does not correspond to the curve of the increasing shear rate. Here we observe a hysteresis. An unconfirmed theory states that in the matter a regulated structure results so that the different molecules which occur in the liquid organise themselves on their own in a specific geometric order. When such a substance is moved, then the given geometric order is disrupted, and the forces which hold the particles and molecules together are broken down. Consequently, no force acts against the liquid flow anymore, and we can observe a drop in the necessary shear stress for maintaining a given shear rate.

7.2.6.2   Curing

When a plastisol cures at room temperature, the viscosity of the plastisol increases. The increase in viscosity can take place faster or slower, depending on the type of the resin used, the type of plasticiser, and the temperature, etc. It is considered that the speed of increase is faster in the first 24 h, and then decreases. If one continues to stir after the initial 24 h, one notices that the viscosity after stirring is lower than immediately beforehand. The difference between the two measurements is what one calls thixotropy. The remaining increase of viscosity between the final measurement and the viscosity of a freshly made plastisol is what one calls curing.

7.2.7   Elements in the Composition of a Plastisol

As we have already seen, an elementary plastisol is comprised of a PVC dispersion resin, a plasticiser that dissolves the resin and thus produces the liquid form, and a stabilising agent which ensures the heat stability. We will discover that many types of raw materials can be, or respectively, must be mixed into the formula to impart different specific properties to the plastisol. In this chapter we will describe all of the possible ingredients and properties which can be imparted by the same. We will begin with PVC, followed by the plasticiser and the additives, which are no less important, but are only needed in smaller quantities.

7.2.8 PVC Resins for Plastisols

The following main properties are described in a PVC catalogue for plastisol producers:

- The K-value, the viscosity index or the inherent viscosity
- The use in compact products, foams or end products
- The viscosity for different concentrations of plasticisers; in general the DOP
- Some information on particle size
- In some cases, heat stability

If one wants to formulate a plastisol one should first find out which type of resin is to be used. That is why it is necessary to ask a few questions:

- Viscosity: what concentration of plasticiser is necessary?
- Foaming ability: is a well foaming resin necessary?
- K-values: Are strength properties important?
- Particle size: do we need a resin with a fine particle size?
- Heat stability: Does the process require a high degree of heat stability?
- Absorption of water: is absorption of water an important factor (exterior application)?
- Haze formation: Do we need a low hazing product (automobile industry)?

Now we will expand on all of these questions, one after the other.

7.2.8.1 Viscosity

The first thing one should know is the degree of flexibility needed. The degree of flexibility is indicated by the Shore hardness, which is directly related to the concentration of the plasticiser. Some tables on the degree of flexibility, as function of the concentration of the plasticiser, can be found in literature (see the section on "plasticisers"). We already know that the viscosity has a direct relationship to the concentration of liquid (plasticiser). Thus, if we know the required Shore hardness of the product to be manufactured, then the necessary concentration of plasticiser is clear. Now the resins can be chosen, which will provide the viscosity as well as the concentration of plasticiser that are demanded by the process. This is only a very simplified way of looking at the problem. Later, additional adjustments can be made, as we will see in the following section. There are many other factors that influence the viscosity of a plastisol, e.g.:

- The curing
- The type of plasticiser
- The type of stabilising agent
- The addition of, and type of extender
- The type and concentration of pigment

Some corrections can be made in the following way:

- By mixing resins with lower plasticiser absorption with resins with higher plasticiser absorption
- By adding an extender resin
- By adding a viscosity diminisher
- By adding a thixotropic substance

Please remember that we have not mentioned the viscosity of the resin, because the resin itself is a powder.

7.2.8.2 Extender Resin

In addition to the dispersion resin, we have now also mentioned the extender resin as a way of correcting viscosity. In specific, the rheology can be changed by the addition of stones of different sizes. A similar effect can be observed in the area of plastisols. We have already spoken of the determining influence of primary and secondary particle size of resins on the viscosity of the paste.

The extender resin is manufactured by way of suspension polymerisation. The special feature of this resin results from very low absorption of the plasticiser, a small medium particle size between 30 and 40 µm, and a regular particle form. Adding an extender resin lowers and stabilises the viscosity. The drop in viscosity has its biggest influence in shear rates in the range of 100 and 2,000 s^{-1}. The drop in viscosity mainly influences the shear-thickening properties of some specific resins.

7.2.8.3 Foaming Ability

Because of their polymerisation formula, some resins foam less than others. This property is always listed in the catalogue.

7.2.8.4 *K-Value*

The *K-value* is based on the viscosity of a solution of 0.5 g/100 ml PVC in cyclo-hexanone at 25 °C, in accordance with DIN 53726. It actually reflects the molecular weight. Based on the same conditions but with a different calculation, one can obtain the specific viscosity and the viscosity index of ISO R 174-1961 (E). The ASTM A method 1243-58 T or 1243-66 is based on the viscosity of a solution of 0.2 g PVC in cyclohexanone at 30 °C. These are the main expressions used. Other expressions can be found when other dissolvers such as e.g. nitrobenzene and 1,2-dichloroethane are used. We must observe that the viscosity index or the relative viscosity is independent of the viscosity of the plastisol. But when the plastisol is stored, the solvation of the PVC by the plasticiser at a low *K-value* is more evident than at a higher *K-value*.

The K-value plays an important role in the gel formation of the plastisol and the mechanical properties of the end product. It is apparent that a PVC with a low K-value and short molecule chains gels more quickly than a higher PVC with a higher K-value. With sufficient gelling of the product, in an end product the strength properties, the tear resistance and the abrasion resistance can be higher, as long as a PVC with a high K-value is used.

7.2.8.5   Gel Formation

After mixing the resin with the plasticiser and all the other additives required, we are left with a plastisol with a given consistency or viscosity. But the goal of transforming to the end product is to produce a strong, soft product. In order to achieve this result, the plastisol is transformed into its final form in a specific process, which we will look at more closely later on. Then a gelling of the liquid phase takes place, to a solid phase, using a curing process. After cooling, the plasticated PVC component has all the beneficial properties of a vinyl plastic, which we are familiar with. This development is also called the "theory of the lost available space". The swelling of the polymer particles by the absorption of the plasticiser when subjected to heat leaves less and less space for the swollen particles (and diminishing liquid phase) to move, until finally, a solid state is reached.

At higher temperatures the swollen particles are fused together, whereby a homogeneous state is produced; the solid state is only reached after the melted mass has cooled down to the *D*-level.

The solid is then in a brittle condition and only achieves its actual strength properties after full melting. The given temperatures for the different phases can vary depending on the plasticiser used to create the plastisol, and the resin.

This is very important because the plasticiser and the resin used can accelerate or inhibit the gel formation process, and also change the viscosity of the melted mass after fusing. Later on, we will see that it is possible to change the viscosity of the melted mass in order to get the viscosity that is necessary to produce a good chemical foam. When a gas in a viscous product is released, the resulting cell structure depends mainly on the viscous state.

This prompts us to talk about additives and plasticisers in a plastisol formula.

7.2.8.6   Particle Size

A few special applications, such as e.g. very thin coatings with a coat thickness of < 100 μm, which are applied by a doctor knife with a roller profile need resins with small particle sizes. A resin with a maximum particle size of 130 µm can apparently produce stripes if the distance between doctor knife and paper is 100 µm and is limited using a serigraphic 100 "penta" screen.

7.2.8.7   Heat Stability

In some applications such as the rotation or slush moulding methods, undercarriage coatings on the car, leather for dashboards in the car, etc. plastisols with a high degree of heat stability are required. Here, resins with a high degree of heat stability are an advantage.

7.2.8.8   Water Absorption

In the case of outdoor uses such tarpaulins, roof coverings, etc. and even for indoor uses such as abrasion resistant floor coverings, low water absorption is in demand. Here, PVC plays an important role.

7.2.8.9   Haze Formation

Low hazing resins and formulae must be used for all automobile applications. Very strict regulations with respect to PVC resins are anticipated in future. For more on, see DIN standard 75201.

7.3   The Plasticisers

7.3.1   Function

Plasticisers impart a certain fluidity to the PVC dispersion resin. That is why they are an important ingredient, without which the production of plastisol would not be possible in the first place. In addition to the necessary degree of flexibility, the plasticiser can impart special properties to the end product, e.g. bendability, fire resistance, or low migration, as we shall see later when looking at the corresponding properties.

This is how it works:

There are many theories on plastication. They can be summed up as follows:

a) PVC is comprised mainly of an amorphous mass with some crystalline areas.
b) When heated, the electrostatic forces between the PVC chains decrease, which makes it possible for plasticiser molecules to be pushed between the chains.

It is clear that the plasticiser molecule must have a certain degree of polarity in order to ensure that it is held in the chain and is not simply discharged out of the flexible

PVC. Polarity is commonly produced by carbonyl groups $>C = O$ in the plasticiser and, if applicable, by aromatic rings, if available. The alcohol groups in the plasticiser are non-polar and attenuate the electrostatic forces between the PVC chains. This raises the flexibility of the PVC. Based on this, one would expect that the length and the linearity of the alcohol chains would have a considerable influence on the performance characteristics. In fact, performance characteristics are a function of linearity and molecular weight. Linearity depends on both the alcohol used as well as the acid used. The linearity of the alcohol varies from the mainly branched OXO alcohols (isooctanol to isodecanol or tridecanol), through semi-linear (approx. 75 %) alcohols such as linevol 79, right to fully linear alcohols which are produced through ethylene growth, such as e.g. alfol 610 and 810.

7.3.2   What are Plasticisers Made of?

Plasticisers are mainly produced by esterification of alcohols with acids or anhydrides. The alcohols used in the manufacture of plasticisers vary between C4 and C13. These alcohols can be attained from crude oil as well as from materials with a vegetable base. Various dibasic aromatic and aliphatic acids or anhydrides are used in the esterification process.

7.3.3   Their Performance Characteristics

Without going into detail about their special characteristics, the main ones are the following:

- Efficiency
- Compatibility
- Volatility
- Solvation, gel formation and fusion
- Relative density

Other characteristics which can be achieved through the choice of plasticiser are:

- Resistance to water extraction, soap, and benzine
- Resistance to hydrolysis
- Low migration
- Resistance to heat, light and weather
- Good bendability
- Fire resistance
- No colouration, no odour and no toxicity

7.3.3.1   Efficiency

The efficiency corresponds to the required concentration needed to achieve a certain degree of flexibility, whereby the DOP serves as reference. This efficiency is expressed by plus or minus in the number of parts plasticiser in one hundred parts PVC "phr": That is the phr value of the DOP required to achieve a degree of flexibility "X".

7.3.3.2 Compatibility

In a mixture with a PVC resin the plasticiser's ability is demonstrated in terms of forming a homogenous mixture. This mixture must be resistant to certain environmental conditions which could lead to an excretion of the plasticiser. In fact, the compatibility is dependent on the number and kind of the functional groups of the plasticiser, as well as the size and form of its molecules. Other influences on compatibility include other additives, secondary plasticisers and chlorinated paraffin, stabilizing agents and the type of PVC (emulsifier).

7.3.3.3 Solvation, Gel Formation and Fusion

The solvation of the PVC brought about by the plasticiser and its compatibility are closely related to one another. Even at room temperature BBP, DBP and TCP have considerable solvation strength. Compared to the less compatible DOS, they can fuse PVC even at low temperatures. This property makes it possible to advance or delay gel formation and fusing by choosing the right plasticiser.

7.3.3.4 Volatility

It is important to do away with plasticisers with low volatility. In the curing process the volatility leads to a loss of plasticiser, which is not very economical or ecological. Even at room temperature the end product loses its flexibility. In the case of technical applications in the automobile industry, a dashboard film on the windscreen in sunlight can sustain an increase in temperature on the surface of up to 103 °C. At night, the temperature of the windscreen can drop drastically. The plasticiser condenses on the windscreen. This phenomenon is called haze formation. Strict requirements are now applied within the automobile industry as a result.

7.3.3.5 Relative Density

The quantity price is strongly influenced by this property. When designing storage equipment it can also be important whether or not free water sinks in the tank or rises to the top.

After this, we will look at the classification of the various plasticisers and the respective, specific properties that are provided by using these plasticisers.

7.3.4   Classification

As already mentioned, diverse alcohols and acid anhydrides are used in the production of plasticisers. This leads to a classification of plasticisers in a certain number of chemical families, as for example phthalates, phosphates, and adipates. Another, more general classification divides the plasticisers as follows:

- Primary plasticisers
- Secondary or special plasticisers
- Diluents

To sum up, we can say that the approach to classification of plasticisers in this way leads to two different interpretations, provided that they do not differ too strongly from one another.

The First Interpretation   In this interpretation the terms primary, secondary and diluting have the following meanings:

- Primary plasticisers: compatible up to 120 phr
- Secondary plasticisers: less compatible. They impart special properties.
- Diluent: hardly compatible. These are never used alone.

The Second Interpretation   The primary plasticiser is the main component used in the formula.

The secondary plasticiser is the other component in the formula with a special function. It imparts special properties.

The difference between these two classifications is that the so-called secondary plasticiser (according to the second version) can be seen as the primary plasticiser (according to the first interpretation). Since the second interpretation does not contain any specification (as for example, compatibility) of a specific plasticiser, the description of the plasticisers in the following table takes the first classification into consideration of the primary and secondary classes.

7.3.4.1   Primary Plasticisers

As follows, we will classify the plasticisers as primary and secondary plasticisers and group them in families.

Phthalates   Because of their performance and price, these are used the most in the coating industry. The properties they contain are dependent on the chain length of the alcohol molecules and their structure (linear/branched). In this group, in particular, we can also find DOP (dioctyl phthalate), which is the best plasticiser in terms of compatibility, efficiency and volatility, measured by the ratio of quality to cost. That is why it is viewed worldwide as a reference for all plasticisers. The plasticisers useful for the plastisols are taken from C4 to C13. The C13 (ditridecyl phthalate) is viewed as the upper limit of compatibility. If we go from C4 to C13, the following applies:

- The efficiency of the softening and compatibility decrease.
- At the same hardness the cold resistance properties rise.
- Volatility decreases.
- Water resistance increases.
- Resistance to hydrocarbons decreases.

The following applies to the structure (linear/branched): An increase in branching, e.g. in the case of DINP, means:

- Reduced cold resistance properties
- Shorter fusing time
- Higher viscosity
- Higher efficiency in degree of flexibility

DBF—Dibutyl Phthalate

- High volatility
- Strong gel formation strength
- Faster curing
- Increased compatibility of a chlorinated paraffin
- Lower elongation distortion
- Higher density
- Is never used alone

BBP—Butyl Benzyl Phthalate

- Volatile
- Quick gel formation
- Quick curing
- Good stain resistance
- Induced shear-thickening in the plastisol viscosity
- High density
- Poor weather resistance
- Low bendability
- Not susceptible to stains

DHP—Dihexyl Phthalate

- Volatile
- Tends towards quick fusing
- Curing
- Useful bendability
- High density
- Known in Europe as Jayflex 77 Exxon

BOP—Butyl Octyl Phthalate In Europe, BOP is known as the plasticiser 216 Monsanto. Numerous different products with different mixtures go under the name of BOP. In order to evaluate, it must be known what is to be coated.

DOP, also DEHP—Dioctyl Phthalate—Diethylhexyl Phthalate As already mentioned, this is the best plasticiser in terms of quality per cost, good compatibility, efficiency, volatility compromise, and low cost. It is no wonder it was chosen as the reference for plasticisers.

DIOP—Diisooctyl Phthalate DIOP can be used as substitute for DOP. It produces a lower viscosity than DOP. Because of the branching it is suspected that this plasticiser is mutagenic and is less biodegradable. That is why DIOP has vanished from the European market.

DINP—Diisononyl Phthalate

- Lower curing of the paste
- Always good gel formation speed

In certain economic contexts, DOP and DINP can both have the same price. DINP is preferred over DOP because of its lower volatility. A study by Dr. Popp of the Exxon company found that even with price differences, from an economical point of view DINP can replace DOP because of its lower volatility and lower density.

Some forms contain an antioxidant (e.g. bisphenol A) which produces a pink tinge in certain foam preparations.

DIDP—Diisodecyl Phthalate

- Less efficient
- Low volatility
- Low curing of the paste
- Low sensitivity to water

In order to ensure that the same Shore hardness of A80 is achieved using DOP, we need in addition 4 phr DINP and 6 phr DIDP. DINP and DIDP are less efficient than DOP.

When using DINP and DIDP, in order to achieve the same gel formation as with DOP, an increase in temperature of 3–6 K is necessary.

DUP—Diundecyl Phthalate DUP is suitable for special applications which demand low volatility and also water extraction or absorption.

DTDP—Ditridecyl Phthalate

- Very low efficiency and volatility
- Low extraction

DTDP is used in the area of plastisols, as plasticiser which is compatible with longer molecule chains.

Alfol Phthalates in C6-C8-C10 and C8-C10 Compared to DOP, a long-chain alcohol phthalate offers:

- Higher efficiency
- Lower volatility
- Good cold resistance, which can be achieved by mixing in DOP/DOA in a ratio of 70/30
- Low viscosity of the paste
- Little hazing

Alfol phthalates are expensive to acquire.

Linevol Phthalate in C7-C9 and C9-C11 Similar properties are produced as with alfol and linevol. The difference between C7-C9 is larger than the difference between an alfol C6-C10 and C8-C11.

Phosphates The main property of phosphates is that they are flame retardant.

In flame retardant plasticised formulae in which the PVC is self extinguishing, the plasticiser is the main problem. For example, a non-elastic PVC has a chlorine content of 57 % and an oxygen index of roughly 37. If this is plasticised with 60 phr DOP, the total chlorine content drops to 36 % and the oxygen index to 22. That is why plasticisers with a phosphate base are used. While some flame retardant products are based on other raw materials such as Sb_2O_3, Martinal (al trihydrates) or other chlorinated paraffins, for example, are used either individually or together with phosphate plasticisers, or other materials, to achieve a synergy effect. Other flame retardants than phosphate plasticisers provide layers which are impervious to light. However, for transparent layers only plasticisers can be used.

The flame retardant properties are relevant for many areas, such as:

- Wall and floor coverings in public buildings
- Air ducts and conveyer belts in the mining industry
- Imitation leather for upholstered furniture in public buildings
- Imitation leather and tarpaulins in the automobile industry
- Public transport (train, ship, bus, aircraft)

The general structure of phosphate plasticisers is:

```
Rl-O                          Aryl              Alkyl                Alkyl
R2-O   P = O   R can be   Aryl P = O   or   Alkyl P = O   or   Alkyl P = O
R3-O                          Aryl              Alkyl                Alkyl
```

Consequently three groups exist:

TAP—Triaryl Phosphate In the past, TCP (Tricresyl phosphate), TPP (Triphenyl phosphate) and CDP (Cresol diphenyl phosphate) were used. But the ortho isomer in the cresylic acid reduces the use. In contrast, TPP has a freezing point of 48.3 °C and tends to crystallise out of PVC. Today, new phosphates are produced from phenols with a high molecular weight. This also includes those which are based on *p-tert*-butylphenol and mixed isopropyl phenols. Their flame retardant properties are comparable to those of TCP and CDP. However, they have poor heat properties. In fact the flame retardant properties of a phosphate plasticiser are dependent on the aromatic compound content, while alkyl content, in contrast, is a disadvantage. However, the properties of the alkyl group assure higher efficiency and better low-temperature characteristics. Sometimes this can be viewed as useful compromise.

TOF or TOP—Trialkyl Phosphates These are successful because of their low-temperature characteristics, not because of their low flame retardant properties.

Alkyl Diaryl Phosphates As can be deduced from what has already been written, this is a compromise between low-temperature characteristics and good flame retardant properties. Depending on the test method and the formula, the alkyl diaryl can provide better or poorer flame retardant properties or smoke emissions than TAP. One of these plasticiser classes, the 2-ethylhexyldiphenyl phosphate, has been regulated by the FDA for use in packaging and in adhesive coatings.

Since phosphate plasticisers are hard to recognise by their trade names, we provide you with a list with a classification of the plasticisers.

The Citrates Often used are acetyl tributyl citrate Citroflex A4 from Pfizer, and Estaflex ATC. Their biggest advantage is their lack of odour or taste, as well as the fact that for a long time they were the only plasticisers that were certified by law, e.g. by the FDA, for food or medical use.

7.3.4.2 Secondary or Special Plasticisers

This class includes plasticisers which, in terms of price and low compatibility, are used exclusively for certain properties of the plastisol.

The following can be found in this category:

- "Low-temperature" plasticisers
- "High-temperature" plasticisers
- Plasticisers with only minimal extraction and migration behaviour

Plasticisers that Provide Low-Temperature Resistance These plasticisers are by-products of aliphatic alcohol of C4–C10 and dibasic acids:

- Adipates (Butane dicarboxylic acid)
- Azelates (Heptane dicarboxylic acid)
- Sebacates (Octane dicarboxylic acid)

For economic reasons, from this group dioctyl adipate (DOA) is used more frequently. Today, special applications demand a lower level of volatility. There are other adipates with longer chains, such as e.g. DINA or DIDA, on the market.

Applicable in cases of increasing chain length of the alcohol or the dibasic acid is:

- A decrease in low-temperature characteristics
- A drop in volatility
- A drop in gel formation strength and efficiency
- A rise in price

They are always used with a primary plasticiser in a ratio that depends on the degree of cold to be achieved.

7.4   Stabilising Agents, Accelerators and Co-Stabilising Agents

As already mentioned, a stabilising agent is used to achieve a smaller plastisol formula. We recall that PVC is a powder and thus needs a liquid component—a stabilising agent—to produce a paste or a plastisol. Moreover, the plastisol must undergo one or more steps of heat treatment before it arrives at the final product stage. In the case of chemically foamed plastisols, the stabilising agent can be either fully or partially replaced by an accelerator. This will be explained in the section on "accelerators".

During this operation, the plastisol must withstand temperatures of between 140 and 210 °C. Here, because of the high temperatures, the PVC can break down and release some hydrogen chloride (HCl), with the formation of double bonds. The number of double bonds, which represent chromophore groups, increases, and reveals itself as coloration ranging between a yellowing to an orange and red if not sufficiently stabilised, and black when fully decomposed. This decomposition takes place faster with oxygen present. The released HCl catalyses this decomposition.

Thus, the main tasks of a stabilising agent should be:

- To capture the HCl released during decomposition, without causing any harmful side effects to occur to the end product or the continued heat stability
- To hang together, itself, with the remaining double bonds in order to strengthen the weakened PVC structure
- To protect the PVC against attack by oxygen and UV light

Furthermore, it should not have any side effects on the properties of the PVC, e.g. the transparency, the colour, the foaming substance, and the toxicity.

This would be the ideal stabilising agent, but in practice suppliers offer a wide range of compromises, because this ideal product simply does not exist. Each of these suggested stabilising agents has its properties, and it is the manufacturers' responsibility to choose these in such a way that the desired properties are achieved for the respective application. Naturally, in the area of plastisols, liquid stabilising agents should be used. In a few special applications in which, for example, ZnO or calcium stearate is required as powder, before the actual mixing takes place, a preliminary mixture (*master batch*) should be made, comprised of stabilising agent and plasticiser.

7.4.1   Background

At the beginning the main stabilising agents were lead compounds.
The main ones were:

- Trivalent bases, tribasic lead sulphate: 3 PbO, $PbSO_4$, H_2O very effective
- Dyphos, dibasic lead phosphate: 2 PbO, $PbHPO_3$, 1/2 H_2O good light stabiliser
- Liquid lead stabilisers dissolved in solvents at a high boiling point are less effective than the others.

They are all poisonous, lead to light impermeability, react with sulphur, and some are very good accelerators.

Barium/cadmium and barium/cadmium/zinc were also used. The liquid stabilising agents which were used for plastisols are comprised of salts of short-chain or cyclic acids with organic phosphite groups and a measure of diluent. The latter is necessary for reducing the viscosity of the compound. Many of them are also given some Zn to ensure resistance to sulphur deposits.

Stabilising agents are crystal clear and have good light resistant properties. Some of them are good accelerators. Barium is for long-term stabilisation, cadmium and zinc were used for short-term stabilisation.

In Europe, their use is prohibited because new health regulations do not permit compounds with heavy metals (cadmium).

Currently the following stabilising agents are left: tin, cadmium/zinc, calcium stearate, and barium/zinc stabilising agents.

The tin stabilising agents are the strongest thermal stabilising agents for PVC, but also the most expensive.

They can be categorised in two groups:

- Mercaptides (contain sulphur)
- Carboxylates (sulphur-free)

In these two groups one can find butyl or octyl derivatives.

7.4.2   Mercaptide Group

Example: dibutyltin thioglycolate or dibutyltin thiobenzoate have outstanding thermal resistance, outstanding transparency, an unpleasant odour and poor light stability.

Example: Di-*n*-octyltin thioglycolate

$$\text{Oktyl S} \diagdown \quad \diagup -CH_2\text{-COO-ethylhexyl}$$
$$\text{Sn}$$
$$\text{Oktyl} \diagup \quad \diagdown S\text{-}CH_2\text{-COO-ethylhexyl}$$

7.4.3   Carboxyl Group

Example: Tin maleate or laurate. These have a lower heat effect but better light stability and a better odour.
   Example: Dibutyltin dilaurate

$$\begin{array}{ccc}
C_4H_9 & & OOC\text{-}C_{11}H_{23} \\
 & \diagdown\;\diagup & \\
 & Sn & \\
 & \diagup\;\diagdown & \\
C_4H_9 & & OOC\text{-}C_{11}H_{23}
\end{array}$$

These represent the basic elements of the tin stabilising agents. Meanwhile, on the market, tin stabilising agents are offered that sometimes represent a more technically complex mixture, or require careful mixing during the process, in order to increase light stability in the one stabilising agent, for example, or in the other, to reduce the odour of mercaptide.

7.4.4   Barium/Zinc Stabilising Agents

After stabilising agents containing cadmium were prohibited (Ba/Cd and Ba/Cd/Zn), the barium/zinc stabilising agents came onto the market. Great progress has been made since then. They have come to be used in practically the majority of applications. These are salts of aliphatic or aromatic acids or complex compounds comprised of different salts, which are mixed together with an amount of diluent and organic phosphites. As we will see later on, some synergies can be found when one uses them together with epoxy plasticisers or co-stabilising agents (epoxidised soybean oil or with a stabilising agent of another fatty acid).

7.4.5   Calcium/Zinc Stabilising Agents

The calcium/zinc group has the same structure as the barium/zinc stabilising agents. However, they are not such strong heat stabilisers. While they were formerly used mainly for applications that came in contact with food, such as plastisols or crown caps, because of the most recent advances, today we find them in applications of a few top coats for cushioning vinyl floor coverings and even in mixtures for applications in the automobile industry (dashboards, side walls, etc.), and because of their good compatibility with isocyanates (polyisocyanurate bonding agents), for the coating of PES and PA fibres. Some of these are offered under their chemical name, such as e.g. zinc oxide, calcium stearate, or zinc octoate.

7.4.6   Accelerators

In chemically foamed plastisols, accelerators based on similar molecules replace partially or almost fully the stabilising agent. A few years ago dibasic lead phosphites and barium/cadmium/zinc were used. In the USA, barium-cadmium-zinc is still manufactured, and in Asia it is still in use. In Europe, these are currently being replaced by barium/zinc, calcium/zinc, potassium/zinc, magnesium/aluminium/zinc, zinc oxide or zinc octoate, but this list is not exhaustive. In fact, in addition to its task as stabilising agent, the accelerator is also used to reduce the thermal decomposition temperature of the foaming agent, which is usually of the type azodicarbonamide. The task here is to decompose this foaming agent at the optimum melt viscosity in order to ensure a fine and regular foam structure. Later we will see how this works.

7.4.7   Conclusions

An important rule in the area of plastisols is the use of liquid stabilising agents and accelerators. These can be distributed easily in the plasticiser. If ever a solid powder is needed, before its use one must first make a master batch. One can turn it into a paste form if one mixes it, if applicable, with a plasticiser and a dispersing agent.

In the case of stabilising agents, one must know that all stabilising agents have specific properties and contraindications. It is imperative that their influence on the following points be known:

- The viscosity of the plastisol
- The so-called pre-stabilised PVC resin
- The transparency
- The original colour
- The long-term durability
- The light stability
- The intensive yellowing or reddening
- The effectiveness
- The haze formation properties
- The reaction with polyisocyanurates
- The compatibility
- The toxicity
- The physiological properties

For accelerators, they must have an influence on:

- The physiological properties
- The reaction with isocyanates
- The haze formation properties
- Inhibitors
- The rheology of the plastisol
- The speed of the accelerator (very fast, fast, medium, slow or very slow)

All of these properties must be carefully read in the manufacturer's catalogue, or discussed with a technician. Even for specialists, ongoing education doesn't hurt. Sometimes useful ideas can even come up in routine conversation.

7.4.8   Co-stabilising Agents

In literature, some co-stabilising agents that are rarely used can be found such as polyalcohols, beta diketones and aminocrotonates. On the other hand alkyl or aryl phosphites such as diphenyl decyl phosphite (DPDP) and trinonyl phenyl phosphite (TNPP) are already integrated to a large extent in commercial stabilising agents. This section deals with the widely used family of epoxidised oils. Synergies have been found with respect to heat resistance when these co-stabilising agents are used together with Ba/Zn or Ca/Zn stabilising agents. In some books, they are described in the plasticiser section as secondary plasticisers, possibly because they are used in great quantities in the USA. It is clear, in formulae in which 5–10 phr are used, that these quantities must be included in the full plastication process so that the Shore hardness of the end product can be predicted. They are also called epoxy stabilising agents.

7.4.9   Epoxy Stabilising Agents or Plasticisers

They are characterised by their oxirane oxygen ring, which is formed by an epoxidation of olefin double bonds.

$$-CH-CH-$$

7.4.10   Epoxidised Soybean Oil

Epoxidised soybean oil is the best known oil with a vegetable origin. It imparts good heat and light stability, and can be used in contact with food. Because of its high molecular weight, it has good oil and solvent extraction properties, and low volatility. It is also known as ESBO. The oxirane number is 6.8.

7.4.11   Epoxidised Linseed Oil

This has a higher oxirane oxygen index, namely 9.0. It has better resistance properties than ESBO in terms of hydrocarbon and mineral oil extraction, but its extraction behaviour in soap water is worse.

7.4.12 *Octyl Epoxy Tallate and Isooctyl Epoxy Tallate*

Because of the lower oxirane oxygen index these are less effective than ESBO. As plasticiser they are more effective than ESBO and impart to the plastisol some better low-temperature properties and a lower viscosity than ESBO. They are also more volatile.

7.5 Foaming or Expanding Agents

After dealing with accelerators in the previous section, and addressing the drop in decomposition temperature of a foaming agent resulting from the addition of an accelerator to the formula, we would now like to look at the subject of foaming. To foam a plastisol there are three possibilities:

- Chemical foaming
- Mechanical foaming
- That which we can call physical foaming

7.5.1 *Chemical Foaming*

Chemical foaming agents are additives that break down into the following when subjected to heat:

- Nitrogen (62 weight %)
- Carbon monoxide (35 weight %)
- Ammonia (3 weight %)

Depending on the type of the foaming agent up to 120–220 ml/g gas is released at temperatures between 140 and 220 °C. This is a large range, but it is necessary. The trick is to decompose the foaming agent at the right moment, by choosing the right foaming agent and the accelerator, and the right amount (phr) of these. This means that the release of gas should take place when the polymer has reached the right melt viscosity. Let us look at what the temperature of gas release first changes:

- The kind of foaming agent used
- The particle size of the foaming substance
- The type of stabilising agent or accelerator
- The quantity of accelerator
- In general the activation of an AZDM (foaming agent of the type azodicar-bonamide) is raised in the base range and consequently inhibited in the acid range.

If an amount of air is blown into the water through an AG4 filter (sintered glass), then large air bubbles result because the air blown in does not meet any resistance. After replacing the water with oils of different viscosities, in the given viscosity small, regular bubbles should remain. If the melt viscosity of a plastisol is too low, then one may observe large bubbles. When the optimum melting viscosity has been reached, one will see small, regular bubbles. If the gas is generated while the melt viscosity is too low, then one gets a foam with a low expansion factor, a rough or irregular surface, and a poor foam structure. Let us now look at how a few additives influence the melt viscosity:

7.5.2   PVC Resin

PVC resins differ from one another because of their alkalinity or acidity, their particle size distribution, the type of emulsifier and other additives, as well as their *K-value* which represents the molecular weight. As demonstrated above, a basic milieu promotes the activation of AZDM. At the same K-value, the resins are easier to melt the finer they are. On condition that the emulsifier is not impure because it contains other polymerisation additives, the type of emulsifier is decisive for a good foam structure. This way resins with the same K-values can result in different foam properties because of the different polymerisation formulae. A resin with a low *K-value* has a low melting temperature and a low viscosity of the hot melt. Nevertheless, for some applications for thermoforming of partially foamed foils, a resin with a high *K-value* is required.

Usually one obtains a closed cell structure; however some resins result in an open cell structure and some of them even produce whiter foam than other, typical quality grades.

Where very low melting temperatures are needed, the copolymers VC-VAC (vinyl-acrylic) can be used alone or in a mixture with homopolymers. The higher the VAC content, the lower the melting temperature (the VAC content varies generally between 3 and 15 %).

7.5.3   Plasticisers

Plasticisers also have a considerable effect on the melting temperature. Compared with DOP, a polar plasticiser such as BBP (butyl benzyl phthalate) is the fastest in producing the melted mass. Nevertheless, a prescribed plasticiser's side effects on other properties must be taken into consideration. Should a BBP plasticiser be chosen to improve the processing by accelerating the melting operation, and to impart good pigmentation properties to the end product, then side effects such as a decrease in viscosity stability during storage of the plastisol, a negative effect on the cold modulus of rupture, the higher specific weight of this plasticiser, and last but not least the economy must all be taken into consideration.

7.5.4 Accelerators

The decomposition of a foaming agent such as azodicarbonamide begins between 210 and 215 °C. Since temperatures between 170 and 205 °C are necessary in a typical foaming operation, one uses accelerators. Accelerators are metal soaps made of barium, calcium, zinc, aluminium, potassium or magnesium.

Zinc oxide is also often used. Practically all kickers used in normal quantities also work as stabilising agents, which means an additional stabilising agent is not required. With azodicarbonamide, without accelerator the decomposition curve begins at roughly 200 °C; at 215 °C it is complete. If an accelerator is added, the decomposition curve is shifted towards the lower temperature range, depending on the quantity and type of the accelerator added. With that, one can condition the decomposition temperature to the temperature at which the optimum melt viscosity is reached. Naturally, this is the theory. In practice it is difficult to measure these temperatures. The formulator of active agent preparations knows:

- Which PVC is to be used; about the *K-value*, the application and other factors if known (emulsifier, etc.)
- Which plasticiser(s) will be used as well as the type and amount
- About the expansion required
- About the desired range of production parameters (temperature, throughput speed)
- About other desired properties: Particularly white foam, not much formation of streaks, etc.

He/she should choose the right accelerator (quick, medium, slow); if necessary, also with other properties. Manufacturers offer a wide range of accelerators, with a description of their main properties. If a predetermined foaming agent quantity has been chosen, in the laboratory three concentrations (phr, % of resin) are determined and tested. If they do not meet expectations, then it becomes easier to recognise the direction in which the amount or the type of foaming agent and/or kicker must each be changed. Granted, a certain amount of experience is beneficial to this search.

7.5.5 Foaming Agents

In reality, in the coating process only compounds from the family of azodicarbonamides and sulfohydrazides are used, and in very special cases, hydroceroles. Theoretically, we can also mention AZDN (2,2′-azobisisobutyronitrile) and NTA (N, N′-dinitroso-N, N′-dimethylterephthalamide) and, for the sake of completeness, some non-organic compounds such as sodium bicarbonate or sodium borohydride.

7.5.6 Azodicarbonamide

This is often abbreviated as AC or AZDM (the most common names). Its exact name is 1, 1'-azobisformamide, or ABFA. All accelerators can be viewed in relation to this foaming agent. One property is that a basic milieu promotes the activation of AZDM. We will see in the chapter on floor coverings that activation is inhibited under acidic conditions. Since the powder has a yellow-orange colour, sometimes it is especially easy to identify a foam-coated plastisol. After decomposition is complete, with the generation of 220 ml/g gas, the remains are colourless and non-toxic.

Some producers offer AZDM in different particle sizes. This can be useful in order to achieve slower decomposition. Between the AZDMs with fine and with large particle sizes, an increase in the exposure time of 30 % can be observed. It has been demonstrated that an AZDM with a fine particle size decomposes more quickly than an AZDM with a coarse particle size. The medium particle size of the finer sort lies between 3 and 5 μm.

The important thing is that the foaming agent is fully distributed in the plastisol. One can imagine that, in the case of an agglomerate, no regular foam structure can be achieved. In the past, at a time when finer sorts were not available, a few so-called foam regulators were used, such as e.g. Galoryl PL386 (Compagnie Francaise de Produits Industriels) and VS 103 (Air Products). Since then, AZDM has become fine enough so that it can be added directly as powder in the mixer. But, in general, master batches are still used. The AZDM powder is mixed alone or together with an accelerator in powder form (ZnO), with TiO_2, with a plasticiser or with some epoxidised soybean oil. The mass is then ground in a three-roller mill and homogenised. If the powder is given directly into the mixer then this approach can be avoided. On the other hand, some AZDM sorts are available in powder form in which particles are covered with a dispersing agent. Common sorts of AZDM are: Porofor ADCM/C1 (Bayer), Luvopor (Lehmann & Voss), Genitron ACSP4 (Bayer) etc.

7.5.6.1 Ready-to-Use Compounds

Applicable to Floor Coverings or Hard Leather Products In this case, the AZDM powder is coated with an accelerator. This determines the ratio of foaming agent to accelerator. But since it is mainly used as decoration foam for floor coverings, it is suitable for inhibition and adapted for a standard formula. The great advantage lies in the fact that it can be added in a powder form. Since the accelerator coats the AZDM particles, activation runs quickly and fully. As a result of this, the foam is whiter and one saves somewhat in the area of the necessary titanium dioxide. The amount one needs to produce a foam is, roughly calculated, the sum total of accelerator weight and the AZDM in phr (percentage of resin). Examples of use are: Genitron SCE (Bayer) or Luvopor 365 (Lehmann & Voss).

The Following Applies to Foaming at Lower Temperatures and for the Back Side Coat of Carpets The principle is the same as above, but the type of kicker and its quantity are calculated in such a way that decomposition at lower temperature is aimed for (at about 160 °C). One also obtains very white, self-dispersible foams. Application examples are: Genitron LE (Bayer) or Luvopor LUV/N (Lehmann & Voss).

7.5.7 *Sulfonhydrazide*

Included here are 4,4′-Oxybis(benzenesulfonyl hydrazide) or respectively OBSH (OB). The amount of gas released is 125 ml/g. It is a white powder whose decomposition begins at 130 °C and is complete at 150 °C. The residue is not poisonous; it has no odour and is white. For economic reasons, it is only used if an AZDM coated with kicker is not perfectly adapted, and in the lower range of melting temperatures, for example in the case of finishes for PVC coverings and in the case of foam coating of temperature-sensitive synthetic fibres in the carpet sector. Since no kickers are required, additional stabilisation must be provided. Application example: Genitron OB (Bayer).

7.5.8 *Hydrocerole BIF (Böhringer Ingelheim)*

This is an inorganic foaming agent. At a proportion of 4 phr, a foam density of 0.4 results, with a temperature of 190 °C and an expansion time of 90 s. The advantages are: one needs no accelerator, the foam is very white and has a very low level of streak formation.

7.5.8.1 Foam Properties

The properties of the foam:

Cell structure	Regular
Cell size form	(Spherical, oval) open or closed
Surface	Smooth, rough
Colour	
Expansion factor	
Foam density	
Expansion speed	

Regularity In laboratory equipment, foams of the same thickness are produced with a rough to fine cell structure (five samples). The foam to be analysed is compared with the scale and classified using a number from 1 to 5. The classification is made with the help of a large, illuminated magnifying glass (magnification 10).

Cell Measurement　A picture is taken under a microscope using a Polaroid camera and a known magnification. A standardised square frame is placed on the photo, and the number of cells in the square is counted. The analysis is performed by comparison.

Form

- Round or oval
- Open or closed cells

Principle　A sample with known dimensions is weighed in standard conditions. After submerging in water the foam is taken out, the outer surface is roughly dried with a towel, and then it is weighed. The weight of the absorbed water, converted to volume, is compared with the dry volume. The latter corresponds to the volume calculated from the external measurements of the sample, less the previously coated volume before expansion.

Surface Roughness　Again, a few reference samples are produced and the surface of the foams are compared and recorded.

Colour　The yellow-index is measured using a calorimeter. Some resins or formulae produce whiter foam than others. On the other hand, unwanted colours can also occur (for example pink or brown) if a DINP stabilised with bisphenol A reacts with an intermediate AZDM decomposition product by colouring, whose intensity and colour depends on the metal ions of the accelerator. No coloration is evident in plasticisers stabilised using the antioxidant Topanol CA. However, Topanol CA is more expensive.

Expansion factor　The expansion factor corresponds to:

$$\frac{\text{Density of the unfoamed material}}{\text{Foam density}} = \text{Expansion factor of the foam}$$

Foam Density　The foam density corresponds to:

$$\text{Foam density} = \frac{\text{Density of the unfoamed material}}{\text{Expansion factor}}$$

Expansion Speed or Ratio　The expansion speed or the ratio corresponds to:

$$\text{Expansion speed} = \frac{\text{Foam thickness}}{\text{Applied thickness}}$$

7.5.9　Practical Aspects

Briefly stated, the following applies: with a given specific weight of plastisol, one wants to obtain a foam thickness that can be expressed in terms of the expansion speed, the foam density, or of a specified expansion factor. When all of the desired properties of the produced foam and the desired process parameters (temperature and expansion time) are satisfied, the formulator creates the formula in terms of PVC, softener, foaming agent, extender and pigment. The calculated weight per square metre of plastisol is spread on a substrate and run through the oven heated to the required temperature. During curing the foaming agent, activated by the kicker, releases its gas and foams the coating until the determined expansion time has been reached. After coming out of the oven, the foam is cooled. The expansion time is decisive for the speed of the system. In fact, the length of the oven is the bottleneck of the system. Without a doubt a product manager strives to reduce this to a minimum. If the desired foam thickness is not achieved, then the following actions must be taken:

- Change the temperature
- Change the expansion time
- Change the foaming agent concentration (phr)

The influence of temperature, expansion time and foaming agent concentration (phr) can be described as follows:

- When the temperature is lowered, the expansion time and the foaming agent concentration must be raised.
- When the temperature is raised, the expansion time and the AZDM concentration must be lowered.
- When the duration is decreased, the temperature and the AZDM content must be increased.
- When the duration is increased, the temperature and the AZDM content must be decreased.
- When the AZDM content is increased, the temperature must be lowered and the duration decreased.
- When the AZDM content is lowered, the temperature must be raised and the duration increased.

In the case of white or natural coloured plastisols, the yellow colour originating from the AZDM is easy to observe. If the foam has a yellow tinge after expansion, this is an indication that not all the foaming agent has decomposed. A good formula is one where a maximum in decomposition of the foaming agent is achieved. Ideally, a specified thickness of foam in a sensible time (economically) and at an appropriate temperature should be achieved, without leaving behind residue of undecomposed foaming agent.

If these conditions are not achieved, then the formula must be adapted. For those without experience, this could mean working out x number of formulae. The best of these must each be changed until the desired result is achieved. No more than one, or at the most two, parameters should be changed at once.

7.5.10 Mechanical Foams

In contrast with chemical foams, where an unfoamed plastisol is coated, and where the foaming process takes place in a curing oven, here the plastisol is foamed at room temperature on a special piece of equipment and coated as foam before it gels. The plastisol (5–6 Pa s) is fed into the funnel of a special mixer and mixed using air. The plastisol is strongly beaten in a turbine mixer with the generation of a slight amount of heat, which is why a cooling device is recommended in order to prevent too early gelling of the PVC on the surface. The foam is stabilised with a surface-active substance in order to prevent further breakdown of the cells. Two different surface-active substances are used:

- A hydrophilic soap as surface-active substance with a closed cell structure. This must be compatible with the PVC resin.
- A hydrophobic silicone as surface-active substance with an open cell structure.

The surface-active substance with soap base has the advantage that one must not add any stabilising agent to the mass, because the type of soap used has a stabilising effect. When one uses a surface-active substance with a silicon base, then a stabilising agent must be added as well. Not all PVC resins produce mechanical foams of the same quality. The type of plasticiser is also important. Sometimes a resin can even produce acceptable foam that is stabilised by a surface-active substance with a silicon base, but with soap there is no foam. That is why extreme care must be taken when choosing the PVC. The properties of the PVC resins are not always indicated. The foams produced using both of the surface-active substances demonstrate a very smooth surface; the surfaces of chemical foams, on the contrary, can be rougher and smoothing is often necessary. It must be considered that the coat of a mechanically beaten plastisol enters the oven as foam. Since foam has heat insulation properties, a curing time must be planned. Formulae with copolymer resins and with quick gelling plasticisers are useful in some cases. In the case of chemical foams the foam is foamed in the oven, subject to exothermic decomposition of the azodicarbonamide. In the case of mechanical foams as back side coat, foams with densities of 0.3 and 0.5 g/ml can be achieved, whereby with chemical foam, on the other hand, foams with a density of 0.25 g/ml can be achieved. The foaming that occurs during the melting of the plastisol in the chemical process guarantees only a minimum of mechanical properties. In the case of mechanical foaming, the temperatures can be raised, but care must be taken so that the foam cells do not collapse in on themselves.

7.5.11 Surface-active Substances

The main producers are: Byk Chemie, Dow Corning and Wacker.

7.5.12 Mixers for Mechanical Foams

Worth mentioning are: Hansa Mixer, Mondomix or Oakes. Basic formula:

PVC, homo or copolymer, or mixed	60	100
Diluents	40	–
Plasticisers or mixed with BBP	70	100
Extenders (careful with respect to the viscosity)	–	50
Surface-active substance (depending on the type)	4	5

7.5.13   Physical Foams

While they are widely used in the USA in applications concerning uplift, their use in Europe is limited to a few producers of thick foams. Nitrogen, carbon dioxide, volatile liquids (trichlorfluormethane, chlorinated hydrocarbons, etc.) are released in plastisol (with or without pressure). The gas is released when heated under pressure. This gas escapes by releasing the pressure. Depending on the process, one obtains an open or closed cell structure.

After we have established the basic formulae for compact applications and foam applications, we should now take a look at the numerous types of raw materials which are generally used to impart specific properties to the plastisols, or to reduce their price. In some formulae, extenders are used in large amounts.

7.6   Extenders

We should divide the extenders into three categories:

- Carbonate extenders
- Carbonate-free extenders
- Additives with an extender function

7.6.1   Carbonate Extenders

7.6.1.1   Natural Origins

These extenders are normally called calcium carbonate and are mostly used in the plastisol area. The other compounds used are used in much smaller amounts. Lime stone deposits can often be found on the surface of the earth.

Calcspar This mineral consists mostly of calcium carbonate. Apart from quartz it is the most common mineral on earth. Lime stone, lime and marble are some of these massive rocks. They are the most frequent form this mineral takes.

Lime Stone This is a very frequently occurring sedimentary rock that consists of up to 50 % *calcspar* ($CaCO_3$). Often found together with lime stone are: *dolomite and aragonite*. Nevertheless, some deposits contain up to 97 % $CaCO_3$. That is high enough to use it directly as industrial raw material.

Dolomite A familiar mineral is $CaMg(CO_3)_2$. When treated with sulphuric acid it forms $CaSO_4$ and $MgSO_2$.

Aragonite This is a mineral variety of calcium carbonate $CaCO_3$ that crystallises in prismatic form.

Chalk This is a form of lime stone.

Marble Marble is the compact and crystalline form of a metamorphic lime stone.

These types of extenders, which are different because of their composition and deposits, produce different properties because of the raw material.

The end product can be examined chemically in terms of:

- The degree of whiteness, dependent on the chemical composition and impurities (Fe)
- The refractive index
- The absorption of DOP or oil, dependent on porosity
- The surface treatment. Some quality grades can be purchased treated or untreated. In the case of treated quality grades, the particles are coated (typically with calcium stearate).
- The particle size distribution, dependent on the milling
- The particle form
- The pH level (important for some formulae)
- The specific weight (for calculating the price)
- The water content, dependent on the drying process

For plastisols, some of these properties are more important than others. The most important are:

Viscosity The viscosity is mainly influenced by the DOP absorption, the surface treatment, and the particle size distribution.

DOP or Oil Absorption, Surface Treatment When a large proportion of the plasticiser is absorbed by the extender, then it is clear that the free, liquid volume is reduced, in which the PVC and other solid particles can move. This causes the viscosity to increase. Compared to the same untreated quality grades, some surface-treated (commonly coated with calcium stearate) extender quality grades avoid this problem to some extent.

Particle Size Distribution The main extender quality grades are offered in different particle sizes. Commonly natural quality grades do not go below a medium particle size of 2 μm in order to ensure efficient processing. In fact, the viscosity and the price are higher, the finer the particle size. In PVC resins the extenders comply with the Mooney equation, which describes the following condition:

$$\mathrm{Ln}\left(\frac{\eta}{\eta_1}\right) = \frac{(K_E V_f)}{(l - V_f/P_f)}$$

η Corresponds to the viscosity of the mixture
η_1 Corresponds to the viscosity of the suspending medium

V_f Corresponds to the volume fraction of extender and PVC resin
P_f Corresponds to the packing component of extenders and PVC resin
K_E Is the Einstein coefficient

Thus it is theoretically possible to regulate the viscosity of plastisols with a high proportion of extender by mixing in extenders with different particle sizes. A comparison: it is known that, in the manufacture of concrete, the flow characteristics are mainly influenced by mixing in pebbles of different sizes.

The refractive index This gives an indication of the light resistance of the plastisol.

Degree of Whiteness We will deal with this property in the section "Application".

Moisture Content The presence of water can produce bubbles in different layers and lead to side effects with other additives, e.g. isocyanates.

Specific Weight This is important for price calculations.

7.6.2 Precipitated Calcium Carbonates

Pure lime stone is subjected to different treatments:

$$CaCO_3 \text{ calcination above } 900\,°C \rightarrow CaO + CO_2{*}$$

$$CaO + H_2O \rightarrow Ca(OH)_2$$

$$Ca(OH)_2, \text{ charged with } CO_2{*} \rightarrow CaCO_3 + H_2O{\downarrow} \text{ (PCC)}$$

It is available in various kinds of crystallisation and degrees of refinement. The medium particle size can be as fine as 7 µm. It imparts to the coat a good light resistance (but less than TiO_2), an excellent degree of whiteness, and has a considerable influence on the viscosity. Precipitated calcium carbonates are offered with or without surface treatment. The influence on the viscosity is dealt with in the section on shear thickening substances and applications.

7.6.3 Carbonate-free Extenders

These are used in the plastisol area to a lesser degree; at best to impart special properties.

7.6.3.1 Silicates

All silicates are minerals that are produced by different methods. They represent the largest class of carbonate-free extenders.

7.6.3.2 Clays

These originate from sedimentary rock and are a result of the decomposition of silicate-containing stone such as gneiss, slate, and granite.

7.6.3.3 Kaolin

Calcinated kaolin improves electrical resistance.

7.6.3.4 Porcelain Clay

This is used in large amounts in the cable industry.

7.6.3.5 Feldspar

Feldspar is used only to a limited extent in the plastisol area as non-sticking agent. The transparency is retained in numbers of parts of up to 7 phr.

7.6.3.6 Barium and Calcium Sulphate

Barium Sulphate Because of its high density ($4.35 \, g/cm^3$), it is quite expensive. It is used in cases where drapability properties must be increased, or a particularly level surface is important (e.g. the back side of a carpet). Barium sulphate is very clear and imparts light resistance against X-rays.

Calcium Sulphate (Water-free) This extender is used in acid-resistant food package foils.

7.6.4 Additives with an Extender Function

Many additives which are used for very special purposes are solid powders and, as extender, can have unwanted side-effects such as an increase in viscosity, or a lessening of mechanical properties, etc.

7.6.4.1 Aluminium Hydroxide, Antimony Trioxide or Antinomy Pentoxide, Magnesium, Molybdenum, Zinc Compounds

These are used as flame retardants and smoke suppressants. Refer to the section on "flame retardants".

7.6.4.2   Pyrogenic Silica, Bentonite, Precipitated Silica

These are used as thickening agents. Refer to the section "Rheologic modifiers as additives".

7.6.4.3   Titanium Dioxide, Soot, Pigments

These are used as colorants. TiO_2 can be used in quantities of up to 10 phr; soot and other pigments, on the other hand, only in small amounts. However, they can have a marked influence on the rheology. Refer to the section "Colour", also for metallic powders.

7.6.4.4   Metal Powders

Metal powders are used for electrical properties or for a metallic effect on the colours.

7.6.5   General Influences of Extenders

The main effects of an extender on a plastisol or other end products are brought about by the particle size and distribution, the packing density, the surface, the presence or absence of a coating, the amount used, and the plasticiser content.

For PCC extenders, the particle size of extenders used in plastisols generally lies between 0.07 and 25 μm, but the normal range is around 5–15 μm. Exceptions occur in the case of the back sides of carpets, where a coarser quality grade can be used. It is generally recognised that, with respect to mechanical properties, extender contents of up to 10 phr have no noticeable effect on the end product. The following properties change for medium-size extender concentrations:

In plastisols:

- The viscosity

In end products:

- The costs (decreasing)
- The light resistance (increasing)
- The tensile strength
- The elongation
- The hardness
- The low-temperature characteristics
- The electrical insulation, with the exception of porcelain clay

These are general reference values, since the influence on these characteristics depends on the type of extender and its properties, for example in the case of electrical insulation.

Until now, we have seen that the type of the PVC resin, the plasticiser and the amount of plasticiser, the type of stabilising agent, and the type and amount of extender can all have a noticeable effect on the viscosity. On the other hand, we have seen that the viscosity of a plastisol is of great importance for the processing. What happens when careful preparation of a formula leads to an unsuitable viscosity preventing processing? In the following section, we will see that the viscosity can be adjusted with the help of a few rheologic modifiers.

7.7 Rheologic Modifiers

Plastisols must be formulated in such a way that they can provide the performance required by the end product. At the same time, the economics must always be kept in mind. Moreover, care must be taken to ensure that the method's requirements, gel formation and viscosity are met. In order to do this, it is initially possible, with some restrictions, to experiment with the selection of raw materials. The first changes would achieve:

In the case of PVC: the *K-value* would be similar. Foamed or compact quality can be found in the area of low, medium or high viscosity. Also possible is a combination of these two kinds: those which strongly absorb plasticisers, and those which absorb plasticisers only a little. By varying the quantities of both kinds, a large viscosity range can be covered. The addition of a diluent resin is possible.

In the case of plasticiser: different kinds can be found with the same possibilities or combinations.

In the case of the stabilising agent: some of these can be used to reduce viscosity. To be recommended is the use of a liquid instead of a solid.

In the case of the extender: the type, the medium particle size, the oil absorption, whether it is coated or not, as well as the choice of a combination of two particle size qualities can all change.

If changes must still be made to viscosity even after all these changes, then the help of rheologic modifiers is necessary. Consequently, these are divided into two groups:

* The thickening agents
* The agents used to reduce viscosity

7.7.1 The Thickening Agents

These are used to increase the quantity of material, to achieve thixotropy or only to increase viscosity. The main thickening agents currently in use are:

7.7.1.1 Precipitated Calcium Carbonates (PCC)

The possible small medium particle size, the large surface as well as the oil absorption (if not surface treated) permit a noticeable increase in viscosity of a plastisol with low shear rate and, to a lesser extent, an increase in viscosity at a higher shear rate. They impart a pseudo-plastic behaviour to the plastisol, and increase light resistance, but to a lesser extent than titanium dioxide. They are generally used in concentrations of 2–20 phr, but other concentrations are also possible. A frequently used quality grade for plastisol has a medium particle size of 0.07 μm and a surface of 20 m^2/g.

7.7.1.2 Pyrogenic Silica

Plastisol quality grades are comprised of ball-shaped primary particles of roughly 10 nm medium particle size, which have a surface of 200–300 m^2/g. The silanol groups found on the particle surface are connected by hydrogen bonds. This produces a three-dimensional network that increases viscosity. When stirred or in the case of a quicker movement of the plastisol such as that which occurs when coating using a doctor knife—i.e. the plastisol is moved under a doctor knife at a specific speed— the network is broken and viscosity drops. The more intensive the movement or the longer it lasts, the more the viscosity drops. If one leaves the plastisol at rest again, then the three-dimensional network forms again. This process is known as thixotropy. The thickening and thixotropy effect depends on the polarity of the system. The best systems are the non-polar ones. Moreover, the intensity and duration of the stirring operation affects efficiency; in other words, the mixer one uses makes a difference. Namely, the smaller the medium particle size and the larger the surface area, the more the viscosity increases. In comparison with a precipitated calcium carbonate, it is clear that smaller quantities are needed. The medium concentration of pyrogenic silica is 0.3–4 phr.

7.7.1.3 Organic Calcium Thixotropy Agent

This special thickening agent consists of very small calcspar crystals (0.03 μm) that are coated with a calcium sulphonate comprised of long hydrocarbon chains as surface-active substance. An interaction between the PVC resin, the pigments and the hydrocarbon chain is produced. The way the chains are meshed with one another produces a gel structure, or a gel, that can be mechanically broken and after resting, can come together again. Ircogel is the name of the proprietary chemical from the company, Lubrizol. Some qualities are thinned in DOP. The main advantage of this thixotropic agent lies in its liquid form, which means there is no occurrence of light particles which fly away. Moreover it has good run resistance, even in hot conditions, and a lower viscosity at high shear rates. On the other hand, in contrast with powders, it retains, if not even improves upon, transparency. A yellowing that occurs can be suppressed by the addition of a special stabilising agent. The medium concentrations lie between 2 and 6 phr.

7.7.1.4 Modified Urea as Thixotropic Agent

This additive is produced by the company, Byk Chemie. Not much has been written about this additive. It is a liquid which has the same advantages as Ircogel. A high degree of deflection resistance can be observed when heated. The plastisol remains liquid after mixing. When a vacuum is applied, air can be effectively removed, and the gel structure develops only after a certain amount of time. Recommended concentrations lie between 0.5 and 5 phr.

7.7.1.5 Agents to Reduce Viscosity

As mentioned at the beginning of this section, it is possible within the framework of the required properties of the end product to reduce viscosity by choosing the right kind of plasticiser. It is generally known that some plasticisers, e.g. linear 9–11 or DOA, impart low viscosity. However, in comparison with general purpose (GP) plasticisers, such plasticisers are expensive, even when they are used in a mixture. Another solution is the partial substitution of e.g. TXIB (Texanol isobutyrate), a highly volatile compound, for the GP plasticiser. It is also common in some plants to add an amount of GP plasticiser, should it be necessary to adapt the viscosity. Naturally, this causes the softness of the end product to increase. When one experiments with a type of plasticiser, one should always keep the overall efficiency of the plasticiser in mind. The use of a substance to reduce viscosity is used as last resort.

7.7.2 Solvents

It used to be that the formulator could use solvents. These chemicals included solvent benzene, naphthas, volatile aliphatic and naphthenic hydrocarbons or mineral oil hydrocarbons, whereby they are viewed as solvents and diluents. For PVC plastisol this is not the case. If the PVC was dissolved, it would swell and its use would not meet the actual requirement on a substance for reducing viscosity. In Europe, turpentine substitute and low-odour kerosene are predominantly used. These are efficient with respect to lowering viscosity, and they are also economical. However, as a result of emission controls that are becoming increasingly stringent, their use has declined drastically.

7.7.2.1 Hydrocarbons as Diluents

Compared to "solvents", these are preferred.

Linear Paraffins They are very volatile and of limited compatibility.

Alkylated Aromatic Compounds These are more compatible than the linear paraffins.

7.7.2.2   Agents to Reduce Viscosity

Most agents used to reduce viscosity consist of a non-ionic surface-active substance. Examples of this are:

- Fatty acids
- Ethoxylates of univalent alcohols
- Alkylphenols

The negative ions are sodium dioctyl or ditridecyl sulfosuccinate.

At 2 phr and at a low shear rate they can reduce viscosity by 30 %. The viscosity at higher shear rate is also reduced.

When choosing substances to reduce viscosity one must pay attention to side effects, e.g. the influence on:

- Heat resistance
- Light stability
- Coloration
- Electrical resistance
- Water absorption

Furthermore, its degree of effectiveness also depends on the final formula. It is influenced by:

- The type of resin used, which depends on the kind of surface-active substance used in the polymerisation formula
- The surface tension of the plasticiser
- The presence or absence of coated or uncoated extenders

Byk offers substances to reduce viscosity for plastisols with and without extenders.

Some surface-active substances with a silicone base also reduce viscosity and are also more effective in releasing air. When such products are used, one must be careful because these can be responsible for the formation of craters and poor printability.

7.7.3   Substances for Flame Retardation and Smoke Suppression

PVC is itself a flame retardant. However, most organic ingredients of a formula diminish this property, depending on the type of molecule and their concentration. Primarily responsible for the loss of this flame retardation is the plasticiser, which is mostly used in the plastisol area. In the section on plasticisers, we saw that phosphate plasticisers such as chlorinated paraffin are used to produce flame-retardant end products. We will also see that even inorganic additives are used.

7.7.3.1   Flame-Retardant Plasticisers

This is a summary of the section on "plasticisers". Most are phosphate plasticisers and chlorinated paraffins.

Phosphate Plasticisers The main phosphate plasticisers are the triaryl phosphates and the alkyl diaryl phosphates.

Triaryl Phosphates The one used most is TCP (tricresyl phosphate) which is based on a natural raw material. However, because of the ortho isomer in cresylic acid derivates, its importance has decreased. An assortment of TPP (triphenyl phosphate) made of synthetic raw materials is commercially available and widely used.

Alkyldiaryl Phosphates It has been observed that phosphate compounds generally promote strong carbonisation, probably through the formation of phosphoric acids which catalyse carbonisation through dehydrogenation or similar mechanisms. To be very effective, the phosphor must remain in a solid state, whereby the carbon works as an effective barrier to minimise heat transfer from burning into the pyrolysis process. In addition to influencing heat transfer by the formation of carbon, phosphorus changes the pyrolysis mechanism so that less combustion gas is generated.

Depending on the formula and the test, they demonstrate a poorer or better flame-retardant and smoke-suppressant effect than triaryls. However, when one compares the triaryls with the alkyldiaryls, the following can be said:

Compared to the triaryls, the alkyldiaryls are better with respect to:

- Their effectiveness
- Their bendability at low temperatures
- Their efficiency (lower specific weight)

These results originate from formulae with phosphates as sole plasticiser. However, in practice, for reasons of economy or because of other factors, they are always used as component in a plasticiser system. From the point of view of the plastisol, their effectiveness is practically the same as that of BBP. However, in high concentrations problems can occur in storage durability, as well as in an increase in viscosity at high shear rates. Since they are also worse in terms of heat resistance, an additional stabilising system must be planned for as addition to epoxidised plasticisers. Monsanto pointed out that a proportion of 5 % Santicizer 141 raises light stability, whereas, on the other hand, high concentrations of phosphate plasticisers lower light stability. It is difficult to categorise the different kinds of phosphate plasticisers in terms of their flame-retardant effect, because their performance does not depend only on the formula, but also on the special kind of test carried out. The tests most commonly carried out are:

Oxygen index (ASTM 2873): this gives the minimum percentage of oxygen in a mixture of nitrogen and oxygen in order to support the combustion of a sample similar to the burning of a candle.

NBS smoke chamber (ASTM E662): the smoke passes through a beam of light and the density of the smoke is measured.

FMVSS 302 "Federal Motor Vehicle Safety Standard": the sample is arranged horizontally and a flame held nearby for 15 s. The maximum propagation of flame is limited to 101.6 mm/min.

French test with electric burner (NF P 92-503)

French Epiradiateur test (NF 501)

ASTM E 84 (UL723), Steiner tunnel test: this is used in the case of construction materials.

ASTM D 3801: this is a vertical combustion test for electrical parts and equipment.

7.7.3.2 Chlorinated Paraffins

It is believed that halogen compounds in the combustion zone have an effect mainly through the formation of hydrogen halides. These complete the free radical chain reactions associated with the oxidation of combustion gas. Some also increase the carbonisation.

These secondary plasticisers, which contain 42–64 % chlorine, are less compatible than the phthalates. Their compatibility depends on:

- The resin used
- The type of emulsifier
- The compatibility of the primary plasticiser used
- The increase in chlorine content (the higher the better)
- The presence of other secondary plasticisers

Compatibility diagrams are provided by the manufacturers; e.g. IQ, Witco, Dover etc.

With their very low effectiveness, their low price and even with their high specific weight (1.15–1.25 N/m^3), one can achieve very low costs of sale. They have a slight odour and stabilise the viscosity. But the higher the chlorine content, the higher the viscosity. Because of their low effectiveness they impart poor bendability at low temperature. In some formulae, they can partially or fully replace the phosphate plasticisers used in combination with other plasticisers, as well as a portion of the inorganic flame retardants. It is important to remain within the area of compatibility, otherwise printing problems can occur.

7.7.3.3 Inorganic Flame Retardants

In addition to the plasticisers there are numerous inorganic materials that can be used as flame retardants, e.g.:

- Aluminium hydroxide oxide
- Magnesium hydroxide, calcium carbonate or hydroxy carbonate
- Antimony trioxide and pentoxide

Others which work primarily as smoke suppressants are:

- Molybdenum
- Zinc compounds

Aluminium Hydroxide Al(OH)$_3$ We can describe these as coolants. In fact, they bring about a general slowing of the reactions taking place after physical transformations or endothermic decomposition. The power of the flame retardants is based on the following principle:

- The decomposition of $2\,Al(OH)_3 \rightarrow Al_2O_3 + 3\,H_2O$ runs strongly endothermic, roughly 300 kJ/mol aluminium oxide is produced, which protects the plastic material against rapid decomposition. The decomposition temperature is at 204 °C.
- The Al_2O_3 generates a protective layer on the surface that limits further burning.
- The water generated in the above reaction forms a protective gas barrier by dispelling the oxygen.

Tests have shown that there is a synergy with:

- Phosphate plasticisers
- Borates; in particular here, zinc borate

In the case of antimony trioxide, however, a different flame retardant, the only synergy effect is an additive one.

In the case of plastisols, the use of aluminium hydroxide produces a pseudoplastic behaviour that is proportional to the fineness of the chosen quality grade. The precipitated fine quality grades work best when they are dispersed; they are also much more efficient for this. The aluminium hydroxide oxides sometimes have a larger oil absorption capacity than calcium carbonates. Some coated quality grades are available for the plastisol area. The specific weight of aluminium hydroxide is 2.4 N/m^3.

Magnesium Compounds The flame-retardant effect of these compounds is based on the same behaviour as aluminium trihydrates. Compared to aluminium trihydrate, they have one main advantage: They have a much higher decomposition temperature. This gives them a larger margin between decomposition and processing temperature.

Magnesium hydroxide and hydroxy carbonate are produced synthetically from a magnesium oxide salt solution. That makes them more expensive than aluminium hydroxide.

In contrast, *magnesium hydroxy carbonate and magnesium calcium carbonates* are naturally occurring compounds. Very pure mineral deposits of hydromagnesite (magnesium hydroxy carbonate) with huntite (magnesium calcium carbonate) have been found. They are more competitive than aluminium hydroxide. But, as with all raw materials, they must be compared with one another with respect to their special properties and their effectiveness, and not just the price.

Magnesium hydroxide	$Mg(OH)_2$
Magnesium carbonates are	$MgCO_3, MgCO_3, Mg(OH)_2, 3H_2O$
Calcium magnesium carbonates are	$CaCO_3, MgCO_3, 3CaCO_3, MgCO_3$

Magnesium and aluminium hydroxides must be applied in higher concentrations so that they achieve their flame retardant effect; between 20 and 30 phr in plastisols. Both work as smoke suppressants.

Antimony Compounds

Antimony Trioxide All antimony oxides (tri and pentoxide) act as flame retardants. Materials that react with free radicals are responsible for the propagation of flames. Antimony trioxide is used more often. Antimony oxides only function in combination with halogen-containing compounds: PVC, chlorinated paraffin, etc. The mechanism is a gas phase inhibition with respect to the production of antimony halogen products. Because of its high index of refraction and its fine particle size, antimony trioxide has a strong pigmentation effect. As a result, a high concentration of pigment is necessary for glowing colours, which leads to expensive formulae. High tint levels are available, but naturally more expensive. The use of antimony trioxide is limited more to grey, black and pastel colours. With these colours it is possible to omit the white pigment TiO_2, but not for outdoor use because the UV shielding effect is not the same. The level of concentration (up to 7 phr) is much lower than for Al or Mg hydrates. That is why these antimony compounds are used very much in the PVC sector. The majority of antimony comes from China, and is virtually monopolistic, which leads to large fluctuations on the market and, from time to time, unaffordable prices. In some cases, for better transport properties, the consistency of a paste is suggested. At 20 °C the relative density is 5.5 g/cm^3.

Antinomy Pentoxide This is limited to PVC applications. It has a very low specific tinting quality, is much more expensive than antimony trioxide, and is less efficient. The relative density is 3.8 g/cm^3.

Compounds with Antimony Trioxide Base Many producers offer pre-mixed compounds with an antimony trioxide base, with additional zinc, molybdenum, etc. These compounds range from diluted antimony trioxide, additional smoke suppression, to synergetic properties. One must be careful concerning the composition of such mixtures.

7.7.3.4 Other Compounds with Zinc and Molybdenum

Zinc Borate (Zinc Derivatives) The more frequently used zinc borate (to some extent replaces the antimony trioxide) achieves a good level of smoke suppression. One disadvantage is the negative effect on the thermal stability of the plastisol. The relative density is 2.5 g/cm^3.

Zinc Oxide (Zinc Derivatives) This achieves varying results with respect to flame retardant capability. However, it is a smoke suppressant, just like zinc borate. In large quantities, it diminishes the heat stability of the plastisol.

Molybdenum Compounds These are quite expensive compounds and tend to be smoke suppressants rather than flame retardants.

7.7.3.5 Pigments

In addition to aesthetic reasons, the coloration of vinyl products can also be used for the purpose of raising the UV resistance, and weather resistance, etc. Pigments or dyes can be acquired in different forms. For plastisol, it can be a powder form or a pre-mixed form which contains finely distributed pigments with various additives (plasticisers, dispersal substances), which were mixed in a high-speed mixer and ground in a three-roller mill. The choice of a dye should be made with great care, taking into consideration its physical and chemical properties, in order to ensure that these are compatible with the method of processing and the desired properties of the end product. Important factors are:

- Heat stability
- Light stability
- How easily it can be worked/kneaded
- Chemical inertness
- Opacity and transparency
- Its capacity to colour
- Electrical resistance

Before we enter the world of colouring (an unbelievable number of pigments are available) we should touch on two specific colours: white and black. Both are sometimes treated separately in literature. This is the result of their specific properties, which they can impart to the end products.

White Colour

Titanium Dioxide (TiO$_2$) Because of its opacity, its high refraction, the strength of its whiteness, its chemical stability and brightening capacity, the predominance of titanium dioxide as white pigment for plastic is unmatched. It is the most important of all pigments (inorganic and organic) in the PVC sector. Depending on the application process, titanium dioxide is available in two crystalline forms: anatase and rutile. Practically only rutile is used in the preparation of plastisol, because of its light stability. Rutile has a considerably higher index of refraction, and thus tinting strength. Compared to anatase, rutile has a higher relative density (4.4/3.9 g/cm^3) and is somewhat more expensive. Its fine qualities, coated with siloxane or organically treated, demonstrate improved kneadability/workability and lower plastic absorption.

Typical end use concentrations are:

Outdoor use as tarpaulin	8−15 phr
Padding	5−10 phr
Wallpaper	15−25 phr
Floor covering	3−8 phr

Above 10 phr the rise in reflection is very low. For thin coatings, e.g. for blinds, the addition of 0.05 % (in the case of TiO$_2$) very fine aluminium powder or of 0.02 % carbon black raises opacity without an unacceptable reduction in brightness.

In addition to titanium dioxide, other white pigments with low opacity can be found. They are used less often, but can lower the price when added to other formulae. They are:

- **Lead white pigment**, discontinued because of toxicity problems
- **Zinc white pigment**, can also contain lead
- **Lithopone**, used as thinner for bleaching; composed of 30 % zinc sulphide and 70 % barium sulphate
- **Antimony trioxide**, was used as pigment in formulae for flame retardants

Black Pigment (Carbon Black) Carbon black can be used as:

- A simple dye
- UV absorber
- A conducting filler

An astonishing number of carbon black materials for hundreds of applications exist on the market. A suggestion: find a few producers such as Degussa, Cabot or Columbian and ask them for a brochure on carbon black for vinyl plastisols. This is the only way to acquire a list of the right quality levels.

In summary, the following applies: if you use it only as dye, then depending on the desired depth and opacity, the concentrations of 0.5–3 phr are to be used. 2–3 phr are sufficient for UV protection, in order to achieve the necessary opacity. As conducting filler, however, it could be difficult because the plasticiser absorption of carbon black is very high, exactly the same as the dose for an antistatic material (up to 20 phr). So if you use carbon black, please note:

- Mix it well in order to avoid agglomerates. A basic premixed batch can help, but note its plasticiser absorption percentage, otherwise the results may be unexpected.
- For an antistatic formula, with some quality grades at 1 phr, the viscosity can be increased three to four fold.
- It is a cheap form of UV protection if black is accepted as a colour.

All Other Colour Pigments We will see that there are two main pigment families—inorganic and organic.

Both have their peculiarities and are divided into other chemical groups. When choosing some pigments, it is important for the formulator to answer a few questions:

- Which colour is needed? (nuance, brightness, base tone, transparency, etc.)
- What requirements on the end product exist? (light stability, weather resistance, toxicity, colour fastness)
- Is it compatible with the production process? (heat stability)
- What is the dispersion capacity, flotation like?

The general aspects of the product are one thing, but the properties of the end product must be retained. For example, there is no use for a construction foil that does not have a weather-resistant colour. If the colour is not weather-resistant, then gradations may occur, and even worse, the mechanical properties could be diminished, which could lead to complaints (Table 7.1).

Table 7.1 Different families of dyes and pigments

Colour	Organic pigments	Inorganic pigments
White		Titanium dioxide made of anatase or rutile
Lead white		Lithopone
		Zinc sulphide
Red	Monoazo	*Cadmium red*
	Disazo	Iron red
	Basic, water soluble colour	Molybdate red
	Quinacridone	
	Athraquinone	
	Pyrazolochinazolon	
	Diketopyrrolopyrrol	
	Isoindoline	
Yellow	Athraquinone	*Cadmium yellow*
	Monoazo	Iron oxide
	Disazo	*Lead chromate*
	Basic, water soluble colour	Nickel titanate
	Isoindoline	Chrome titanate
	Tetrachlor isoindoline	
	Azomethine	
Orange	Monoazo	Molybdate orange
	Disazo	*Cadmium orange*
	Quinacridone	
	Athraquinone	
	Isoindoline	
	Tetrachlor isoindoline	
	Pyrazolochinazolon	
Brown	Disazo	Iron oxide
	Isoindoline	Zinc ferrite brown
Green	Basic, water soluble colour	Chrome oxide green
	Phthalocyanine green	Cobalt green
	Nitroso	
Blue	Basic, water soluble colour	Cobalt blue
	Phthalocyanine blue	Ultramarine blue
	Athraquinone	
	Alkali blue	
Violet	Calco oil violet	Manganese
	Athraquinone	Ultramarine violet Cobalt violet
	Basic, water soluble colour	Black iron oxide
	Carbon black	
	Aniline	
Black		

For ecological reasons, substances in italics are not used anymore

What are the advantages and disadvantages of the individual groups?

Organic Pigments

Advantages: • Transparency
                • Brightness
                • High tinting strength

Disadvantages: • Susceptible to weather
                • Light stability
                • Heat resistance
                • Tendency to run
                • Dispersability

Inorganic Pigments
Advantages: • Opacity
             • Heat resistance
             • Dispersion ability
             • Light stability
             • Resistance to running
             • Weather resistance

Disadvantages: • Dull/pale colours
                • Low tinting strength

Special Effect Pigments

Mother-of-pearl Mother-of-pearl produces a mother-of-pearl or metallic effect (gold, silver). It is much more reliable than metal flakes, and not reactive with acid, as are bronze metal flakes. One supplier is Merck Darmstadt, for example.

Metal Powder Included here are aluminium, bronze or copper. Associated with fine aluminium powder is a risk of explosion. When it is used as fine powder, it produces a satin effect. When it is used as a coarser powder, it produces an improved metallic effect.

Fluorescent Pigments with a fluorescent effect can be used for a large number of colours, but their light stability outdoors in daylight is poor. In higher concentrations, when one is careful, one can mix it with other pigments; the addition of UV stabilising agents can extend its service life.

Phosphorescent This effect is achieved mainly by zinc sulphide or zinc sulphide in combination with other metals. The degree of brightness is greater in the blue and yellow spectral ranges. Blue glows the longest.

Optical Brightener These are used to overcome the yellow tinge of some soft PVC products, and to increase the brightness of black and white products. However, the long-term effect is not guaranteed. Bayer produces Uvitex OB, $2,2'$-$(2,5$-thiophenediyl)bis(5-*tert*-butylbenzoxazol) very efficiently, but the price of such products remains very high, even though the recommended concentrations are low: 50–100 ppm.

7.8   UV Light Stabilising Agents

Stiff or soft PVC is in itself not a very weather-resistant plastic. After spending some time in the open air, its colour begins to change, it loses colour, it chalks, and loses its mechanical properties; it decomposes. A large range of UV light absorbers exists to protect the PVC from photodegradation. Products that absorb UV light of 295–400 nm are dependent on their structure. Usually the ones most commonly used are hydroxyphenyl benzotriazole and hydroxybenzophenone. The effectiveness of such products is influenced by:

- The concentration
- The thickness of the sample
- And the raw materials present in the mixture

It is apparent that, in the plastisol sector, because of the low specific weight compared to PVC, a large proportion of the volume is represented by the plasticiser, which is why it must be carefully chosen (see section "plasticiser"). Just as important to use, but only in a small concentration, is the heat stabiliser. The use of carboxylate di-alkyl instead of mercaptide (see section on "stabilising agents") is recommended. But not only the type of stabilising agent (good or bad weather resistance) should be considered, but also the efficiency of the heat stability. Also the type and concentration are of the highest importance. If the composition of the plastisol is poorly heat stabilised, or if it must withstand overheating during its production, then the decomposition already begins. Decomposition because of too little heat stability is comparable to being illuminated by UV light. Resistance to UV light and resistance to heat are mutually dependent. That is why the choice of a UV light stabiliser is very much dependent on the heat stabilisation system. Naturally, all other ingredients must also be carefully chosen.

7.8.1 Other Successes in Weather Resistance

As already mentioned, UV light stabilisers absorb the harmful wavelengths of UV light to prevent photodegradation of the PVC. But there is another way (see section on "pigments") to make PVC mixtures resistant to the penetration of UV light—white and black pigments. Quantities of 5–15 phr of rutile titanium dioxide or 2–3 phr of black pigment are sufficient to guarantee good weather resistance. The following applies to the other colours: if the end product is impermeable to light, and if the right coloration can be achieved with a titanium dioxide background of 5 phr plus a concentration of other pigment colours, then perhaps no additional light stabiliser will be used, unless an even higher degree of light stability is desired. Additional colour pigments of the light-stable sort should strictly be used, so that if titanium dioxide cannot be used because of a dye, a cost issue, or simply because the end product must be transparent, another UV light stabiliser can be used as alternative, for a certain degree of light stability.

7.8.2 Important UV Light Stabilisers

In applications containing plasticiser, the UV light stabilisers should be soluble and compatible with the plasticisers used.

7.8.2.1   2-{2′-Hydroxy-3′,5′-(di-*tert*-butyl(or amyl))phenyl}benzotriazole

It is characterised by a high extinction coefficient at 400 nm, with maximum absorption at 340–355 nm. Very efficient are: Tinuvin 320 and 328 nm Ciba Geigy and Cyasorb UV541, a 2-(2′-hydroxy-5′-tert-octyl-phenyl)benzotriazole from American Cyanamide Co.

7.8.2.2   2-Hydroxy-4-*n*-octoxy-(or 4-methoxy)benzophenone

Here, maximum absorption lies generally under 300 nm. It absorbs rays over 400 nm, which can lead to yellowing in higher doses. It costs less than benzotriazole and is still very effective. Chimassorb 81, Cyasorb UV 531 and Uvinul 408 (BASF) can be named here.

7.8.2.3   Sterically Hindered Amines HALS (Hindered Amine Light Stabilisers)

The oligomeric HALS types are soluble and compatible with the main plasticisers. They should not be used alone with tin mercaptide. Included here are Chimassorb 944 EL, Tinuvin 622 LD and Cyasorb UV 3346.

7.9   Adhesion Promoters for Plastisols

In the plastisol sector numerous applications deal with the direct coating or laminating with plastisol of a textile. In both cases an adhesive specification is necessary. Adhesion to rough textile fibres is no problem as long as textiles with natural fibres are used, such as cotton. On the other hand, in the fabrication of synthetic fibres that are made of smooth, endless mono fibres, and to which plastisols adhere only very poorly, a specific adhesive is needed in the formula. This adhesive develops a strong chemical bond at the transition point between fabric and PVC. Two kinds of adhesive are available on the market.

7.9.1   *Single-Component Adhesives*

In the case of this kind of adhesive, a polyisocyanurate is usually dispersed in the plasticiser. The highly reactive isocyanurate group is supposed to react with the polar group of fibres and form networked bonds. But every material in the surrounding area that contains polar groups such as $-NH$, H^+, $-OH$ or $-SH$, should react just as well with the isocyanurate group of the adhesive. All additives that contain such polar groups should be avoided. Vulcabond VP from Akcros, and adhesion promoters 2001 or 2007 from Bayer can be mentioned.

Adhesion promoter 2001 is an aromatic polyisocyanurate in DBF; it is solvent free.

Adhesion promoter 2007 is an aromatic polyisocyanurate in DOP; it is also solvent free.

Vulcabond VP is a polyisocyanurate in DBP.

7.9.2   *Two-Component Adhesives*

This is a mixture of adhesion promoters, for example:

a. TN/50: is a hydroxyl group-containing polyester 4–8 %
b. Desmodur N100: is an aliphatic polyisocyanate 2.1–4.3 %
c. Desmodur N75: is an aliphatic polyisocyanate 2.1–4.3 %
d. Desmodur L75: is an aromatic polyisocyanate in ethyl acetate 2.7–5.4 %
e. Stabilisator 1097 0.2–0.6 %, or
f. VPKA9142 (if an additional extension of the processing time is needed) 0.2–0.6 %.

We will learn more of their use in the section on "tarpaulins".

7.10   Biocides

When a plasticised PVC end product is subjected to a critical environment, high moisture and high temperature, it can experience attack on a microbiological level by bacteria, fungi and algae. While PVC, per se, is unlikely to be attacked by these forms of life, one shouldn't forget that PVC is only one part of a formula. Despite this, it is necessary for us to address E-PVC. Residual emulsifiers are an additional source of carbon. The following end products are susceptible to microbiological attacks: indoor use:

- Shower curtains
- Bathtub mats
- Floor coverings
- Wallpaper
- Water tanks

Outdoor use:

- Tarpaulins
- Tents
- Automobile interior cladding
- Swimming pool covers

Signs of such attacks are:

- Discolouration
- Pink stains
- Yellow stains
- Black stains
- Bad odour
- Loss of mechanical properties, e.g. elongation
- Stickiness

In order to overcome these problems, it is necessary to use a selection of biocides which inhibit or destroy microbiological growth.

7.10.1   Microbiological Attack

Certain conditions, which are essential for all life, are necessary for the growth and strong propagation of micro-organisms:

- Water, direct contact, moisture, high humidity, 60–80 %
- Oxygen (not always anaerobic)
- Temperature
- "Nutrition"
- A certain pH level
- Light

7.10.2   Fungi

Fungi are the most destructive.

They prefer a temperature between 22 and 30 °C and a pH level between 3 and 10.

Nutrition: as we have already seen with the plastisols, plasticisers are present in considerable quantities. What they need is an organic source of carbon. In addition to plasticisers, the stabilising agents and the co-stabilising agents also help. Not all plasticisers are attacked to the same extent:

Very resistant to attack are: DINP (branched), DHP, DIOP, polyester (polymeric) and TXP.
At a medium level of susceptibility to attack are: DOP, DBF.
Very prone to attack are: DOA, SBO.

Many fungi produce dyes or pigments of pink, and yellow to black. A pink colouration is generally produced by Streptoverticillium reticulum, which is present in ground and water. It cannot be removed because it is soluble in the plasticiser. These fungi not only attack PVC but also textiles.

For example: on the fabric edges of boot tarpaulins, because of the capillary effect, water can follow the fibres into the fabric and make black lines along the threads appear transparent (*Aspergilliis niger*) even if the plastisol contains a biocide. But also a simple laminated textile can be attacked and the micro-organisms transferred to the PVC. That is why textiles should also be treated.

7.10.3 Bacteria

They are less destructive than fungi.

Their preferred temperature lies between 25 and 37 °C; the pH level they prefer is neutral to base.

Some bacteria (anaerobic) can live without oxygen.

7.10.4 Algae

The growth factor for algae is photosynthesis.

7.10.5 Common Biocides for Plastisol

Liquid biocides are preferred because of the ease of mixing, but they are not the most stable.

The biocides are usually used in a mixing ratio of up to a maximum of 2.5 weight %.

OBPA 10,10′-Oxybisphenoxarsine For indoor applications, the percentage by weight in the formula lies at up to 1.5 %, and for outdoor applications at up to 2.5 %.

Products which can be mentioned here include Intercide ABF from Harckros or Vinizene from Morton International.

Octhilinone 2-*n*-Octyl-4-isothiazolin-3-1 For these chemicals, for indoor applications the percentage by weight lies at up to 1.0 %, and for outdoor applications at up to 2.5 %.

One product here is e.g. Micro-Check from Ferro.

All these products are poisons and should be handled with care.

Other than chemists who design PVC, biochemistry is not an area of general knowledge, and when the addition of a known biocide is not adequate, it is necessary to explain the problem to the producers of biocides and to ask for a specific solution.

7.11   Other Raw Materials

7.11.1   Antistatic Agents

These are used in areas in which sparks could cause damage; for example:

- Air transport systems or conveyer belts in mining
- Floor or wall coverings in hospitals and computer rooms
- Industrial curtains

A good approach is to first decide on the best possible choice of the main raw material:

- E-PVC with a large proportion of emulsifier
- Active plasticiser (if affordable): the use of a very active plasticiser supports the mobility of antistatic agents; e.g. DOA.
- Large quantities of ESBO (if there are no objections)
- Possibly conductive carbon black (can only be used to a limited extent in the plastisol sector because of its high degree of plasticiser absorption)
- Metallic extenders

If not possible, an antistatic agent must be used. There are different types:

- Polyol ester
- Amines:

    - Ionic, cationic
    - Anionic

7.11.2   Calcium Oxide

In surroundings with a high level of humidity, some raw materials can take on a certain amount of water and be responsible for bubbles and craters. Very fine particles dispersed in the plasticiser comprised of CaO can then be added to the plastisol. The manufacturer must then be careful not to induce side effects with other raw materials. The smallest possible amounts must be used. Producers are: Byk Chemie and Sturgess.

7.11.3   Air Releasing Agents

Air present in plastisols can be responsible for craters, bubbles and clouding or opacity in a transparent application.

The presence of air can also have other sources and causes:

- PVC resin types (mainly dependent on the kind of emulsifier and the quantity)
- Plasticiser (DOA allows for good release of air; phosphates on the other hand don't)

- Stabilising agents
- Viscosity (the lower the better)

That is why it is necessary to first choose the right formula. If additional help is needed, certain additives such as BYK 3155 can be used.

7.12　Economy

While raw materials are generally acquired on the basis of weight, the end products are often sold on the basis of volume; e.g. in a 3 mm thick floor covering. Just as, for example, when laying out floor covering, where the worker adjusts his knife to a specific length in order to lay out a layer with a thickness of x or µm, which is quite a large volume in an area of one square metre. So, when we generally have cost per weight, we should be capable of using the relative density of the material to calculate the cost per volume. This way we should be able to calculate the volume costs of the material. In order to refresh our memories let us look at the definition:

7.12.1　Relative Density

The relative density of a material is the weight of a specific volume of a material at a specific temperature divided by the weight of water with identical volume at the same or at a different temperature. That is why no unit exists for it.

As a reminder: 1 L water weighs 1.0 kg at 3.98 °C; at 20 °C the weight is 0.998 kg. Thus there is hardly a difference. In practice, the relative density at temperatures of 20/20 °C, 25/25 °C or 20/4 °C (1 L H_2O: 1.0 kg at 4 °C) is given. So the values of the relative densities should always be stated with the associated temperatures.

Relative density can also be expressed by the ratio:

$$\text{Rel. density} = \frac{\text{Density of the material at a specific temperature}}{\text{Density of water at the same/a different temperature}}$$

The density is the weight of the material per volume unit at a specific temperature.

7.12.2　Formula Examples and Costs

We will now introduce the term "**kilo per volume**", which is the volume of 1 kg water, for example, at 20 °C.

The "**kilo per volume costs**" are calculated by a correction of the price/kg of an additive, whereby the difference between the relative density of the additive and the

water must be taken into account. The relative density of water at 4 °C is 1,000 kg/m^3.
The "kilo per pound costs" are the same as the cost per weight. The equation is:

$$\text{"kilo per volume costs"} = \text{weight costs} \times \text{relative density}$$

Let us now look at two plasticisers:

Aside from price fluctuations, it is possible that the price of DINP per kg is the same as that of DOP.

When we adopt the above assumption, the following results:

$$\text{DOP kg per volume costs} = 0.92 \,€ \times 0.986 = 0.91 \,€$$

$$\text{DINP kg per volume costs} = 0.92 \,€ \times 0.972 = 0.89 \,€.$$

Now we can see the significance of the previous statement.

If we go back to the section on "plasticisers" we will additionally see that there is a difference in efficiency between these two plasticisers. We shall now examine this subject in more detail.

Let us look at a simple formula that contains the two plasticisers, bit by bit and correspondingly calculated, in order to arrive at a foil with the Shore hardness A of 70. In order to arrive at a Shore hardness A of 70, we need 68.6 phr DOP and 73 phr DINP (Tables 7.2, 7.3, 7.4, 7.5 and 7.6).

Formula			
PVC	100	100	100
DOP	68.6	68.6	
DINP			73
Stabilising agent Sn (tin)	1	1	1
Total kg	*169.6*	*169.6*	*174*

	Rel. density SG 20/20 °C in g/cm^3	Kg divided by rel. density		
PVC	1.4	71.43	71.43	71.43
DOP		0.986	69.57	
DINP (MOBR)		0.972	70.58	75.10
Stabilising agent Sn	1	1	1	1
Total kg per volume		*142*	*143.01*	*147.53*

	€/kg			
PVC	0.87	87	87	87
DOP	0.92	63.11		
DINP		0.92	63.11	67.16
Stabilising agent Sn	2.48	2.48	2.48	2.48
Total €		*152.59*	*152.59*	*156.64*

€/kg (total €/total kg)	0.8997	0.9002	0.9002
€/kg per vol. (total €/total kg per vol.)	1.0745	1.0663	1.0617
Savings in % in formula	1	0.76	1.19
Relative density plastisol (total w/total kg/vol.)	1.194	1.186	1.179

Table 7.2 Rel. density at 20/20 °C of phthalate plasticisers and efficiency for shore hardness A 70 in g/cm^3

DOP Di-2-Ethylhexylphtalate	0.986	1.00 SF
BBP	1.121	0.94
DIOP	0.985	1.00
DCP	0.973	
DINP (MOBR)	0.972	1.06
DINP (SLBR)	0.975	
C9 linear	0.969	
DIDP	0.968	1.10
C7/C9/C11 linear	0.973	0.99
N-C6/C8/C10 610P	0.976	0.99
N-C8/C10 810P	0.971	
DUP	0.954	1.14
Undecyl dodecyl	0.959	
DTDP	0.953	1.27

Table 7.3 Rel. density 20/20 °C of phthalate plasticisers in g/cm^3

DOP Di-2-ethylhexylphtalate	0.986	1.00 SF
DOA	0.927	0.92
DIOA	0.928	0.94
DINA	0.98	
DIDA	0.918	
DOZ	0.918	
DOS	0.915	
TOTM	0.991	1.17
TIOTM	0.991	1.19
Epoxytallate-2-ethylhexyl	0.922	
Epoxidised soybean oil	0.996	
Epoxidised linseed oil	1.034	
Acetyl tributyl citrate	1.05	
TXIB	0.945	

Table 7.4 Rel. density of flame retardant plasticisers in g/cm^3

DOP Di-2-Ethylhexylphtalate	0.986
TCP	1.168
Tri-2-Ethylhexyl phosphate	0.926
2-Ethylhexyl diphenyl phosphate	1.093
Isodecyl diphenyl phosphate	1.072
Isopropyl phenyl phosphate CDP	1.18
Isopropyl phenyl phosphate TCP	1.17
Isopropyl phenyl phosphate TXP	1.16
Chlorinated paraffin	
42 %	1.16
45 %	1.16
52 %	1.25

Table 7.5 Rel. density 20/20 °C of stabilising agents in g/cm^3

Ca/Zn	0.90–1.00
Ba/Zn	1.00–1.25
Sn	1.00–1.15

Table 7.6 Rel. density 20/20 °C of powders in g/cm^3

Extenders	
CaCO$_3$	2.71
Clay (calcinated)	2.68
Barium sulphate	4.47
Mica	2.75
Flame retardants	
Antimony trioxide	5.50
Antinomy pentoxide	3.80
Aluminium hydroxide	2.42
Magnesium hydroxide	2.40
Hydrogenated magnesium carbonate	2.50
Zinc borate	2.50
Molybdenum oxide	4.69
Calcium oxide	3.30

The "substitution factor" (SF) shows the efficiency of the plasticiser. For a Shore hardness of A 80, 52.9 phr DOP is used. Therefore the DINP vs. DOP SF is 1.06.

$$SF = 1.06 \times 52.9 \text{ phr DOP} = \mathbf{56.2 \text{ phr DINP}}$$

Chapter 8
Emission Control in Coating Companies

Given the constantly increasing amount of stress to our environment caused by waste and pollutants in wastewater and exhaust gases and due to the increasingly negative impact on our living conditions and quality of life, environmental conservation continues to gain importance. Large concentrations of air pollution cause damage to humans, animals, plants and buildings. Thus, a different, ecological consciousness is developing in the relevant and responsible persons in industry and politics, as well as in individual citizens (Troge 1985). The handling and removal of solid waste, the processing and cleaning of contaminated wastewater before it is directed into standing or flowing waters, and the disposal of pollutants from exhaust air before it is released into the surrounding atmosphere is increasingly becoming a general interest. Which is why heat recovery also plays an important role. It ensures that accrued heat and thus energy be made reusable to a certain degree. This is not only important from an ecological point of view, but also from an economic one. If recovered energy can be reused, there will be lower overall energy and resource consumption.

In a coating process, solvent blends accumulate that cause exhaust air pollutant levels of 3–7 g/m^3. The exhaust air cleaning process must be designed and integrated into production operation in such a way that on the one hand the requirements in the TA-Luft (German technical instructions on air quality control) on emission values are fulfilled and on the other the highest possible economic feasibility is achieved. In order to provide insight into air quality control possibilities and to make the right choice, all known exhaust air treatment processes, their design, effectiveness and applications will be presented.

8.1 Air Pollution

Air is a colourless, homogenous blend of gases. It is comprised of approx. 99 % oxygen and nitrogen; the remaining percent is divided between noble gases, water vapour and carbon dioxide, see Table 8.1 (Breuer 1987).

A. Giessmann, *Coating Substrates and Textiles,*
DOI 10.1007/978-3-642-29160-9_8, © Springer-Verlag Berlin Heidelberg 2012

Table 8.1 Composition of air

Substances	Vol. (%)	Weight (%)
Nitrogen (N_2)	75.52	78.08
Oxygen (O_2)	23.01	20.95
Argon (Ar)	1.29	0.93
Carbon dioxide (CO_2)	0.04	0.033
Neon (Ne)	0.0012	0.0018
Krypton (Kr)	0.0003	0.00011
Xenon (Xe)	0.00004	0.00009
Hydrogen (H)	0.00005	0.000072

If the composition of air is changed—due to solid, liquid or gaseous substances—then we are dealing with air pollution. These substances may be:

- smoke,
- soot,
- gases and vapours,
- dust and aerosols, and
- odorants.

The vapours may also include water vapour (Henselder 1990). If the substances have harmful effects on the environment, they are called pollutants.

Smoke Smoke is understood to be air that contains airborne particles, the finely disbursed solid or liquid particles which come from combustion processes. The size of the particles, which give smoke a predominantly gray to black colour, ranges between a diameter of 1.6 and 10 µm. As condensation nuclei, the airborne particles are involved in the formation of mist and cloud droplets and precipitation.

Soot Soot enters the air as finely disbursed, mostly speckled carbon (Boisserée 1962).

Gases and Vapours Air and all gaseous substances are gases. Their characteristic trait is that they do not have a specific form and in contrast to solid and liquid substances, do not fill a specific space. However, gases have a specific weight. One can describe them as liquid without a specific surface with very low density.

All gaseous substances that have been formed from liquids through temperature increases and that can return to the liquid form are called vapours. Vapours can also form out of solids (Flury 1969). Many solid substances can be converted into a liquid and subsequently a gaseous state through heat increase, e.g. through the application of heat. In cases of sublimation, i.e. direct conversion from the solid to the gaseous phase, solid substances are converted into vapour without melting. It is also possible to bring vapours into a constant liquid form in contrast to the gases, which cannot exist in liquid form under normal pressure and temperature conditions.

The gases and vapours that accumulate as pollutants in a coating company are primarily organic solvent blends. Said blends are primarily comprised of hydrocarbons (petrol, toluene, benzene), alcohols, esters, ketones and amides.

Table 8.2 Dust classification according to deposit speed

Class	Grain diameter in µm	Deposit speed in m/s
1	>5	0.01
2	5−10	0.01
3	10−50	0.05
4	>50	0.1

Dust and Aerosols Finely pulverized solids that float in the air or in other gases are called dust. In accordance with the VDI guidelines 2119, a distinction is made between three types based on the diameter of the particles:

- Coarse dust $=$ grain diameter > 10 µm
- Fine dust $=$ grain diameter 0.5–10 µm
- Very fine dust $=$ grain diameter < 0.5 µm

Particles with a grain diameter under 1 µm are called aerosols. The smaller and lighter the particles are, the longer they remain in the atmosphere and the greater their involvement in chemical and photochemical processes. In the right weather conditions, air pollution caused by dust and aerosols is visible as a smog dome.

The TA-Luft categorizes dusts with regard to their grain diameter into four different size categories that correspond to the deposit speeds, see Table 8.2.

Dusts and aerosols are mainly formed through combustion or metallurgical processes brought about through heating, industry, traffic and vegetation fires, for example. However, airborne particles can also be formed through the chemical reactions of gaseous components of the atmosphere. In addition to photochemical conversion, the formation of NH_4Cl from NH_3 and HCl or the oxidation of SO_2 to SO_3 and the subsequent formation of H_2SO_4 aerosols also serve as examples (Birkle 1979).

The last possibility for the formation of dusts and aerosols, which from an industrial point of view is insignificant, is provided for the sake of completeness. The substances also make their way into the atmosphere in the form of finely dispersed material that is produced on land or over the ocean through wind and exchange processes.

Odorants The odours found at home and in businesses are generally based on variously composed mixtures of odorants. Whether they are found to be pleasant or repulsive depends on individual taste, which is shaped through the individual's familiarisation, customs and upbringing.

As odorants only occur in very small concentrations—generally below 1 g/m^3-in exhaust gases, they are well below the metrological detection limit. Thus, so-called odour threshold values have been defined with the help of various individual's olfactory impressions. There is also the odour count, which is always given as a multiple of the odour threshold. It indicates the ratio of the volume flow rates when an exhaust gas sample is diluted up to the odour threshold.

Odorants are also characterised by high adsorptivity. This is demonstrated, for example, by the fact that they have a lasting cling to skin, hair, fabrics and clothing.

8.1.1   Emission

Emissions (lat. emittere = emit) are defined as the air pollution emitted by a plant. They are contained in the gases that escape into the air during technical processes, in exhaust gases and in the air discharged from a room. The mass of the emitted pollution based on time is called the pollutant mass flow, measured in kg/h. It is based on the emission that occurs during an operating hour under unfavourable operating conditions and with proper operation of the plant. However, if the mass of the emitted substances is based on the volume of exhaust air, said volume is the pollutant concentration with the units $[g/m^3]$ or $[mg/m^3]$, where the exhaust gas is in a normal state, i.e. at 0 °C and 1,013 mbar. In this respect, the pollutant concentration is also subdivided based on whether or not the moisture content of the water vapour is removed before or after the normal state. The German Federal Emission Control Ordinance is based on the exhaust gas volume in a normal state after the moisture content has been removed (BImSchV 1988).

A third option for indicating emissions is the pollutant mass ratio, which refers to the ratio between the mass of the emitted substances and the mass of the products generated or processed in units [kg/t] or [g/t].

8.1.2   Immissions

Immissions (lat. immittere = immit), or pollutants, are understood to be the effect on flora and fauna at ground level caused by the emission. The measurement is taken at a height of 1–5 m.

Harmful environmental effects can be caused by immissions if based on their type, scale or length they are likely to cause dangers, significant detriment or serious inconvenience to the general public or neighbourhood (BImSchG 1990). The pollutant concentration is also indicated for pollutants, whereby the mass of the air pollutants is based on the volume of polluted air. The units are $[g/m^3]$ or $[mg/m^3]$.

Another pollutant is dust precipitation, a time-related mass coverage that is measured in units $[g/(m^2\,d)]$ or $[g/(m^3\,d)]$.

8.1.3   Threshold Values

A large number of threshold values that may not be exceeded were developed to protect humans, animals, plants, soil, water and material goods. A distinction is made between:

- Substance-related emission threshold values,
- Plant-specific emission threshold values,
- Maximum allowable concentrations (MAC) and
- Maximum pollutant concentrations (MPC).

Fig. 8.1 Air quality control regulations

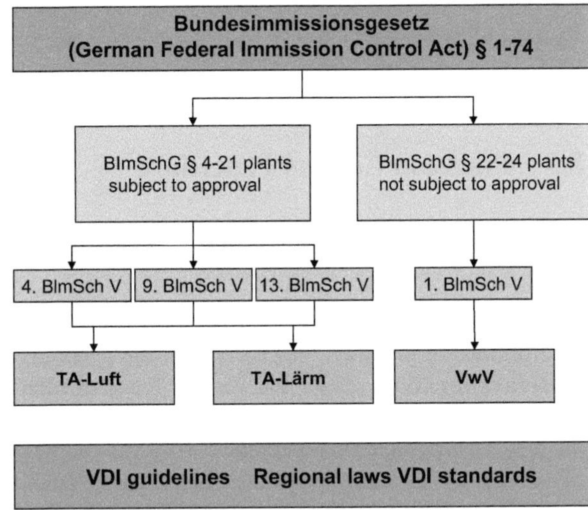

Substance-Related Emission Threshold Values These threshold values are generally applicable to carcinogenic, dusty, gaseous or vaporous, inorganic, organic and odour-intense substances.

Plant-Specific Emission Threshold Values These threshold values apply to furnaces and processing plants in mining, in energy supply, the steel industry, heat supply and the chemical industry.

MAC These values indicate the maximum pollutant concentration that an employee can absorb during an 8-h exposure time without harming his health. The standards for determining MAC values are scientifically substantiated criteria from the health authorities.

MPC The maximum pollutants concentration, which should protect mankind and its environment from harm, is subdivided into the MPCC for constant exposure and MPCS for short-term exposure.

8.1.4   Air Quality Control Regulations

To maintain air quality, there are a series of laws and regulations from the federal government and the state governments. Figure 8.1 shows the most important ones and their relation to one another. Explanations of Fig. 8.1:

BImSchG: German Federal Immission Control Act (Bundesimmissionsschutzgesetz)

BImSchV: German Federal Immission Control Ordinance (Bundesimmissionsschutzverordnung)

TA-Luft: German technical instructions on air quality control
TA-Lärm: German technical instructions on noise
VwV: General administrative regulations

8.2 Emission Control Procedure

8.2.1 Procedure with Solid and Liquid Pollutants

The procedures used for dusts and aerosols are based on their particle size, amongst other things. In this way, the procedures can generally be divided into inertia force separators for coarse dusts and various types of filters and wet scrubbers for fine dusts. All dedusting procedures are based on the principle that relative movement of the dispersed particles vis-à-vis the carrier gas can be induced through the influence of external forces, and the particles then separate from one another. By so doing, the particles reach zones where they can no longer be caught in the gas flow.

There are four main procedures for dedusting exhaust gas or exhaust air:

- Inertia force separators,
- Filtering separators
- Electrical separators and
- Wet separators.

8.2.1.1 Inertia Force Separators

Procedures in which particles are separated from the gas stream through field forces such as gravity, inertia force and centrifugal force are called inertia force separators.

Mode of Operation The inertia force separators include the gravity separator, which is the simplest and cheapest type of separator, and the centrifugal force separator.

The gravity separator reduces the flow velocity through cross-section enlargement. The large dust particles can then be separated due to their increased settling velocity.

Better separation is achieved with centrifugal force separators; separation is assessed with the total separation rate T_{ges}.

$$T_{ges} = \frac{mG}{mA}$$

mG = amount of separated particles
mA = amount of particles fed into the system

or

$$T_{ges} = 1 - \left(\frac{CF}{CA}\right)$$

CF = Cleaned dust content
CA = Untreated dust content.

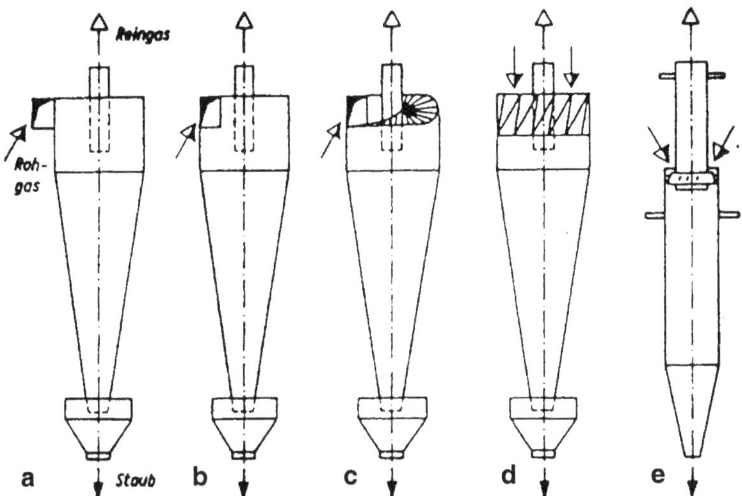

Fig. 8.2 Types of inertia force separators. **a, b, c** tangential cyclone; **d** axial cyclone; **e** multi-cyclone. (Heinrich 1987)

With centrifugal force separators, the polluted air is put into a rotary motion in such a way that the dust particles radially move outward due to the centrifugal forces that occur.

Design The equipment for both separators is comprised of a cylindrical container with a conical lower part. These cyclones—as the apparatuses are called—are available in various models (Fig. 8.2).

The most important component is the immersion pipe. Its diameter determines the pressure loss and thus the possible separation power. The dimensions of the remaining components are adjusted to the immersion pipe, i.e. the cyclone is constructed around the immersion pipe (Kranich 1985).

8.2.1.2 Filtering Separators

The filtering separators are used when highly effective separation is desired, regardless of the fineness of the dust.

The basic principle in this procedure is that the gas to be cleaned is directed through a porous medium, the filter. The filter can be comprised of either fibrous layers (fibre bed filter) or grainy layers (packed bed filter).

Fibre bed filters can be further divided into storage filters and cleaning filters. The main application for storage filters is found in air-conditioning and ventilation technologies, where there is only a low dust content. In contrast, the cleaning filters are primarily used where there are high raw gas dust loads. Because the separation power of the fibre bed filters is much better than that of packed bed filters, the latter are used as pure dust separators when the gas and dust content cannot be handled otherwise. Last but not least, they are excellent for hard, stripping dusts, at high temperatures and with chemically aggressive exhaust gases.

8.2.1.3   Electrical Separators

In this procedure, the solid or liquid particles in a gas are separated with the help of electrical fields. The process in an electrical separator can be subdivided into four consecutive steps (Vans-teenkiste 1978):

1. Charging the particles,
2. Transport of the charged particles to a collecting electrode,
3. Adhesion and the formation of a layer on said electrode, and
4. removal of the dust layer from the collecting electrode.

The direct current field, which is formed by a discharge electrode (as cathode) and a collecting electrode, charges the particles at a voltage of 20–100 kV. Transport occurs through the migration of charged dust particles to the collecting electrode, where they release their charge. A dust layer forms because of the prevalent adhesive forces (electrostatic attraction and van der Waals forces); the layer becomes ever more compressed due to the dynamic pressure of the electrical wind. Said layer must be regularly removed, as otherwise proper adhesion cannot be guaranteed. Electrical separators are subdivided according to the geometric shape of the collecting electrode.

Pipe Electric Filter A pipe electric filter is comprised of multiple pipes that are arranged in parallel and through which gases run either vertically from top to bottom or from bottom to top. Because of the rotationally symmetrical design of the discharge wire and collecting electrode, the electrical field is homogenously distributed and dust accumulation is equally distributed (Fig. 8.3).

Plate Electric Filter In contrast to the pipe electric filter, this filter works with vertically arranged plates. The gases run through the clearance between the plates primarily in a horizontal direction at a speed of 1–4 m/s (Fig. 8.4).

8.2.1.4   Wet Separators

Wet separators—also known as wet scrubbers—put the dispersed particles in the gas flow in contact with a so-called washing fluid. Through this contact, the particles form a dust liquid mixture and can thus be separated from the gas. Wet scrubbers are needed in particular for clingy or highly flammable dusts.

8.2.1.5   Dust Separator Selection Criteria

The type of separator that should be used for specific types of requirements depends on three criteria:

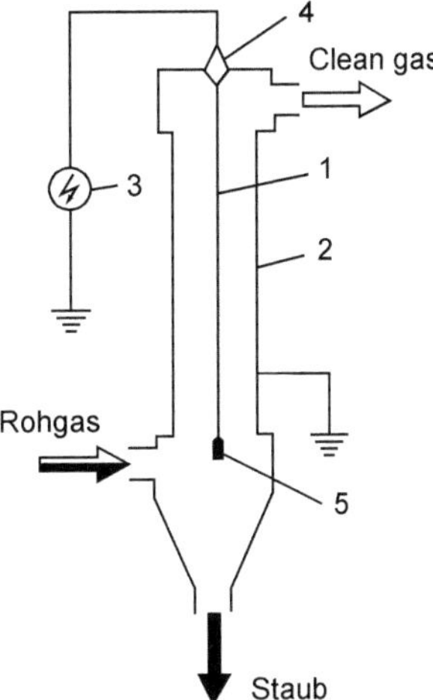

Fig. 8.3 Pipe electric filter.
1 Discharge wire, *2* collecting
electrode, *3* high-voltage unit,
4 isolator, *5* loading weight.
(Fritz 1990)

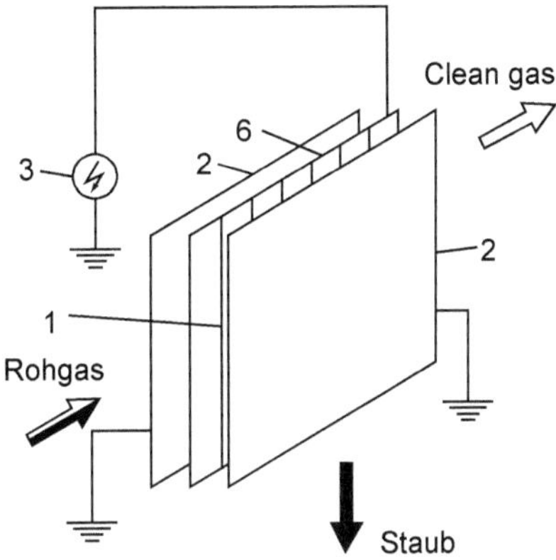

Fig. 8.4 Plate electric filter.
1 Discharge wire, *2* collecting
electrode, *3* high-voltage unit,
6 discharge frame.
(Fritz 1990)

- efficiency,
- total costs, and
- operational safety.

In addition to these central aspects, there are additional criteria that need to be taken into account, such as space requirements, previous experience, etc. The efficiency of the separators described varies greatly. The type of dust to be separated determines which device is best suited to the task. For example, wet separators guarantee the best possible separation in terms of efficiency for highly clingy, pyrophoric, or highly flammable dusts. However, from an economic point of view, the total costs are crucial. The costs are comprised of the annual investment, energy, maintenance and repair costs. Naturally, the costs should be kept as low as possible (minimum principle). The third criterion, operational safety, provides information on the functional safety and on the operational life of the units and their components. They are first and foremost affected by wear (mechanical abrasion) and corrosion.

8.2.2 Procedure for Vaporous and Gaseous Pollutants

There are five procedures to choose from that have proven themselves in the last years with regard to the removal of vaporous and gaseous pollutants (mainly solvents):

- Condensation procedure,
- Absorption procedure,
- Adsorption procedure,
- biological sorption procedure, and
- combustion procedure.

8.2.2.1 Condensation Procedure

In this procedure, the vapour is condensed, which in turns allows for the separation of the substances to be eradicated by cooling a vapour/air mixture below the dew point. Condensation can occur in two ways. With indirect cooling, a flow of exhaust air loaded with vapours passes by a cooling surface. By contrast, with direct cooling the flow of exhaust air loaded with vapours is brought into direct contact with the coolant. The concentration of the vapour, e.g. a solvent, is thereby cooler in an exhaust gas in relation to how much the temperature drops and how high the overall pressure of the vapour/air mixture is. The application of this procedure is limited to high solvent concentrations with small carrier gas concentrations. In some cases, precision cleaning may need to be carried out through adsorption or subsequent combustion.

8.2.2.2 Absorption Procedure

In the thermal adsorption separation procedure, vaporous and gaseous substances are absorbed into a liquid. This means that the concentration of organic compounds in the exhaust gas can be reduced by washing it with a liquid.

As the gas can be absorbed physically or chemically, the procedure is subdivided into physical and chemical absorption. In chemical absorption, the absorbed gas components undergo a loose chemical connection with the absorbent (required washing agent). The absorbent is used up in such a way when it absorbs air pollutants that in most cases it cannot be regenerated. In contrast, regeneration is entirely possible after physical absorption through distillation or purging it with water vapour. Regeneration of the loaded washing agent or solvent, so-called desorption, is significant in terms of economic considerations when purchasing such an apparatus. Unfortunately, combinations of multiple basic unit operations are often necessary.

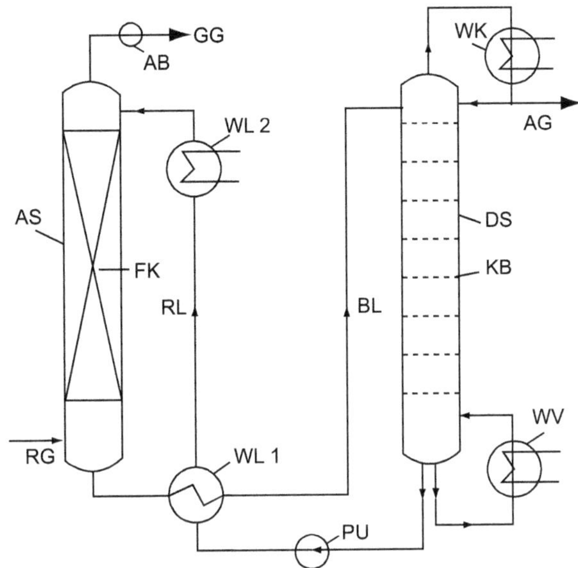

Fig. 8.5 Absorption system with desorber (Sattler 1982). *AS*—Absorber; *DS*—Desorber; *AB*—Liquid separator; *WL1*—Precooler for the regenerated solvent; *WL2*—Aftercooler for the regenerated solvent; *WK*—Capacitor; *WV*—Boiling vessel; *PU*—Pump; *GG*—Cleaned gas; *BL*—Loaded solvent; *RL*—regenerated solvent; *RG*—Raw gas; *AG*—Exhaust gas; *FK*—Packed bed; *KB*—Plate column

Design   In the absorber (Fig. 8.5), the raw gas flowing in is cleansed of its pollutants up to the required or desired purity using the absorbent. Then, the washing agent, which is loaded with pollutants, reaches the desorber, where it is regenerated and subsequently sent back to the absorber after it has been cooled to the required absorption temperature. The height of the absorber is significant; the higher the required degree of absorption (degree of washing) and the lower the ratio of the washing agent flow rate to the gas flow rate, the higher the absorber must be.

Applications   Physical absorption is used for substances that have a lower boiling point than water:

- Methanol,
- Ethanol,
- Acetone,
- Formaldehyde,
- Acetaldehyde, and
- formic acid.

It is also used for some substances with a higher boiling point than water:

- Dimethylformamide,
- Glycols, and
- acetic acid.

For chemical absorption, a distinction is made between the acidic and the base characters of the substances to be eradicated. Acidic pollutants are:

- Phenols,
- Cresols, and
- Thiols.

Basic exhaust gas pollutants are caused by:

- Amines,
- Pyridines, and
- Sulphur compounds.

8.2.2.3 Adsorption Procedure

Adsorption is also a thermal separation procedure. Specific components are absorbed out of the exhaust gas to be cleaned by the surface of surfactant solids. Like absorption, a distinction is made between physical adsorption (physisorption) and chemical adsorption (chemisorption) based on the type of bond. Often, both bond types are present.

Physisorption pertains to a light bond created by van der Waals forces between the surface of the solid and the molecules accumulated there. It can be quickly dissolved.

With chemisorption, there is a chemical bond, the bonding force of which is clearly stronger. A stronger separation effort is necessary to break it.

Regeneration (desorption) is always possible with adsorption. Thus, the adsorbed substances are removed from the surface of the solid and the adsorbent can be reused. A distinction is made between two desorption methods based on the substance system and the chosen working conditions (Menig 1977).

Temperature Change Procedure With this type of desorption, the solid loaded with the pollutants is regenerated using superheated steam or gas.

Pressure Change Procedure The pressure in the adsorber, which is increased after the load phase, is reduced through the pressure change procedure. The gas components adsorbed are released and purged with a flushing gas.

Depending on the type of pollutants to be removed, activated carbon and molecular sieves are mostly used as adsorbents (required solid). Aluminium oxide, bauxite, Fuller's earth, magnesium oxide and silicic acid gel are also used. Generally, only grainy adsorbents with a diameter of 1–8 mm are used. The pores of the substances are in a range below 8–10 cm with large, specific surfaces.

Mode of Operation The exhaust gases loaded with pollutants are pushed through an adsorber filled with adsorbents with a blower. The pollutants are then separated from the exhaust gas via the surface of the solid and the cleaned gas is removed from the adsorber. However, if the exhaust air to be cleaned constantly accumulates, then most of the time a second adsorber unit is required. It is possible to switch to the second adsorber if the first adsorber is saturated, allowing the first to regenerate in the mean time. Such a chain of two or—depending on the amount of exhaust air—more adsorber units allows for continuous operation.

Design In Fig. 8.6, a cooling unit has been turned on ahead of the actual adsorber units. This is necessary if one is dealing with a very oxygen-rich exhaust air, as the exhaust air temperature must then be brought below the carbon flash point.

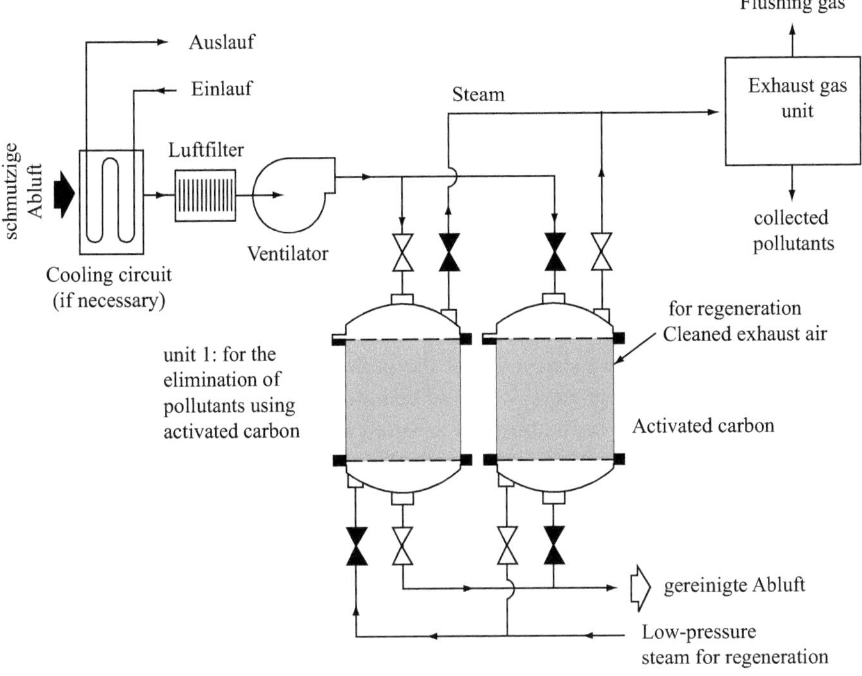

Fig. 8.6 Adsorber with cooling unit. (Peavy 1985)

8.2.2.4   Biological Sorption Procedure

Biological sorption procedures convert the pollutants contained in the exhaust gases into harmless products with the help of microorganisms. The pollutants that can be made harmless with these procedures are primarily odour-intensive substances.

The actual process is always run using O_2, as the reaction equation shows:

$$\text{Pollutant} + O_2 \xrightarrow{\text{Microorganisms}} CO_2 + H_2O + \text{cell substance.}$$

Biological sorption is either done with a bioscrubber or a biofilter.

Bioscrubber Biological decomposition of the pollutants with bioscrubbers can occur in two ways:

- Bioaeration procedure and
- Trickling filter procedure.

In the bioaeration procedure, the aerobes (microorganisms that consume oxygen) floating freely in the water form the so-called activated sludge with the undissolved pollutants. The decomposition speed of these microorganisms is very low; thus, large aeration tanks must be created to regenerate the washing liquid.

In contrast, trickling filter systems are comprised of installations with large, specific surfaces on which the microorganisms are permanently settled. One also hears talk of a biological lawn. The washing water that flows over the lawn supplies the microorganisms with the necessary oxygen and substrates. The washing liquid in which the micro organisms live must also contain nitrogen, phosphorous, and trace elements in order to build up the micro organisms' cell substance (VDI Guidelines 1985). The concentration of the microorganisms should not be more than 10 g of dry substance per litre.

Biofilter In the biofilter, the microorganisms are permanently settled on a carrier like in the bioscrubber's trickling filter systems. Compost, peat, soil or a mixture of these is primarily used as a carrier. The surface of the carrier materials sorbs the pollutants out of the exhaust gases; the pollutants are then decomposed by the microorganisms. In the process, it should be noted that the stock and the activity of said microorganisms is only optimally ensured if the specific milieu conditions in the layer are adhered to (VDI Guidelines 1984). This includes sufficient moisture, the appropriate pH value and oxygen content, and the right temperature. Because the microorganisms are heavily impacted by changes to their milieu or living conditions, long adjustment times must be expected in the event of a change.

There are currently approx. 100 biofilters in use in the Federal Republic of Germany. They are mainly found in purification plants and rendering plants.

8.2.2.5   Combustion Procedure

Combustion procedures convert flammable substances that are responsible for polluting the air into nontoxic components through oxidation. They are also called oxidation processes. The oxidation process is the conversion of fuels to CO_2 and H_2O while supplying oxygen. The general combustion formula is:

$$C_mH_n + \left(m + \frac{n}{4}\right) O_2 \Longrightarrow mCO_2 + \frac{n}{2}H_2O.$$

The following conditions must be fulfilled for combustion:

1. A minimum amount of oxygen must be on hand; most of the time it is taken out of the available air.
2. The required ignition temperature must be reached to be able to start the combustion.
3. With gases and vapours, the mixture ratio with the oxygen must be within the upper and lower ignition limit so that the heat released in the reaction is enough to maintain a minimum temperature.
4. The fuel must be well mixed with the combustion air so that a complete combustion can be carried out.

The required minimum amount of oxygen mentioned in Point 1 can be calculated as follows:

$$O_{min} = \left(\frac{c}{12} + \frac{h}{4} + \frac{s}{32} - \frac{o}{32}\right)$$

c: mol of carbon
s: mol of sulphur
h: mol of hydrogen
o: mol of oxygen.

Because the oxygen enters the combustion chamber with the air, there is a minimum
air requirement of:

$$L_{min} = \frac{O_{min}}{0.21}.$$

Due to the fact that in practice not every oxygen molecule finds a partner in the specified retention period, combustion does not take place completely with the presence
of O_{min}. Thus, more oxygen than the minimum amount of oxygen must be available
for complete combustion. The combustion itself can take place either thermally or
catalytically. Therefore, a distinction is made between:

- Catalytic exhaust air cleaning (CEC) and
- Thermal exhaust air cleaning (TEC).

In older literature one also finds a description of this oxidation process called
"catalytic reburning" (KR) and "thermal reburning" (TR).

Catalytic Exhaust Air Cleaning   In catalytic reburning, oxidation is carried out
using catalysts. The catalyst is responsible for positively influencing the reaction
speed, i.e. ensuring that the speed is increased or the reaction is guided in a certain
direction without actually taking place in the reaction itself. The combustion process
is as follows:

The exhaust gas to be cleaned is heated to the required ignition temperature of
150–300 °C in a combustion chamber either with gas or oil firing. From there, it
goes into the catalytic converter, where it burns without flames on the surface of
said converter at a temperature of 300–500 °C. In the setup shown for a catalytic
afterburning system, the cleaned exhaust gas transfers part of its energy content to the
exhaust gas that still needs to be cleaned in a downstream heat exchanger (Fig. 8.7).

The efficiency of this type of burning essentially depends on the temperature and
the holding time of the exhaust gases in the combustion chamber, on the type and
concentration of the contaminants to be removed, and on the choice of catalytic
converter. The essential process parameters such as temperature and holding time
must be experimentally determined in the process.

The catalytic converter s are mainly comprised of precious metals (platinum,
palladium) on metal supports or metal oxides on ceramic supports (aluminium oxide),
as only highly porous substances with a large, specific surface of some $100\,m^2/g$ of
catalyst mass come into question.

Requirements for Catalytic Converters   There are a series of requirements placed
on the catalytic converters to guarantee economical, operationally reliable, and
environmentally friendly afterburning:

- High selectivity (acceleration of reactions),
- Thermal resistance,

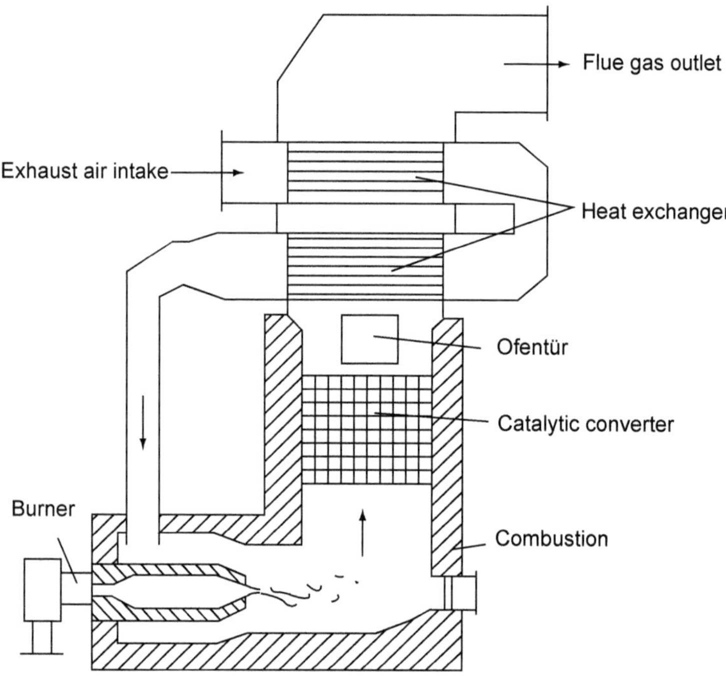

Fig. 8.7 KAR system with heat exchanger. (Baumbach 1990)

- Favourable ratio between surface and volume,
- Chemical resistance,
- Sufficient pressure and abrasion resistance,
- Low pressure loss,
- High service life,
- Low price and
- Cost-effective reprocessing.

The requirements and characteristics listed also serve as factors for the selection of a suitable catalytic converter.

Thermal Emission Control As with catalytic emission control, thermal emission control also requires a mix of fuel and air and sufficient amounts of air and holding time. However, with temperatures of 700–1,000 °C, the reaction temperatures for TAR systems are significantly higher than for KAR systems.

Mode of Operation Organic compounds are cracked at the abovementioned temperatures during thermal emission control because they are thermally unstable. This means that they are broken apart. The carbon and hydrogen released primarily oxidise in the presence of oxygen into CO_2 and H_2O. As the required concentrations are often not available to release enough energy for oxidation without a supply of external energy, the system must normally be supplied with energy in the form of additional fuel. Additional fuels can be heating oil, gas, waste solvents or old oils.

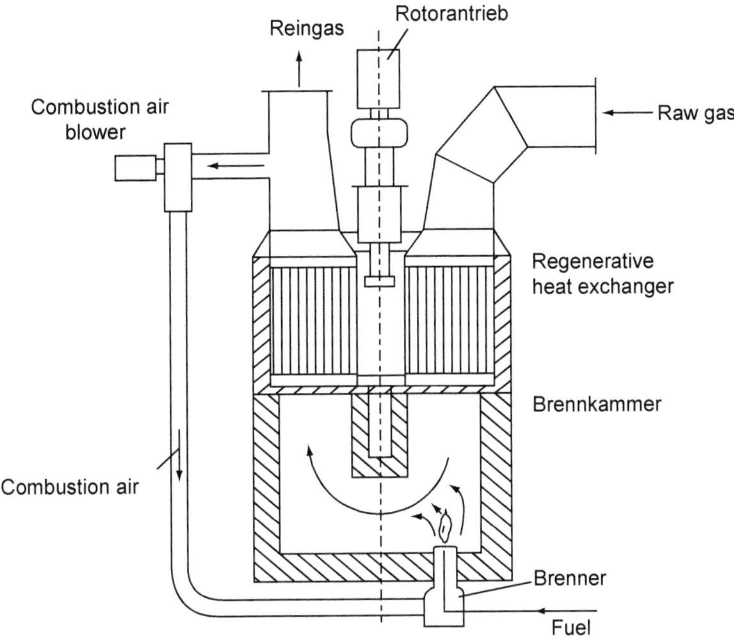

Fig. 8.8 TAR system with heat exchanger. (Baumbach 1990)

The mass flow of these additional fuels is automatically regulated and is based on the supply of contaminants and their heating values as well as on excess air and other substances that enter into the combustion chamber. The use of waste heat cannot be foregone if a lot of additional fuel needs to be supplied, alone for economical reasons. There are three options here:

- Preheating the combustion air,
- Heating the exhaust gases and
- Heating heat carriers.

The procedure in a TAR system looks as follows:

As shown in Fig. 8.8, the clean gas is sent through a regenerative heat exchanger and indirectly preheated by the hot clean gas. Oxidation of the organic compounds (e.g. solvent mixtures) takes place at temperatures of approx. 800 °C. In the process, the burner operates with as little excess air as possible in order to avoid further increasing the gas amounts.

The most important components in a TAR system are the burner, the combustion chamber in which contaminants are burned, and the heat exchanger, which leads to improved system efficiency through the indirect heating of the exhaust gas. The procedural flow chart in Fig. 8.9 illustrates the typical and simplified design of a thermal emission control system.

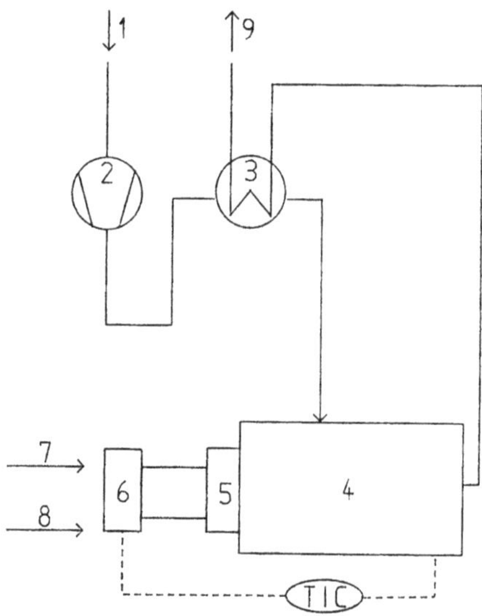

Fig. 8.9 Procedural flow chart for a TAR system, *1* raw gas; *2* exhaust gas blower; *3* heat exchanger; *4* combustion chamber; *5* burner; *6* burn control; *7* additional fuel; *8* combustion air; *9* clean gas; TIC temperature control

The burner's functions are comprised of mixing, heating, igniting and subsequently burning fuel and combustion air. The hot flammable gas flowing out of the burner is used in the process to ignite the exhaust gases that need to be cleaned as evenly as possible so that the contaminates can oxidize. As the type of burner is very decisive in terms of the formation of NO_X—the more excess air available, the more nitrogen atoms can bind into NO—the burner should be operated as near-stoichiometricly as possible. Near-stoichiometric combustion causes the free N atoms to lose the ability to find oxygen as a reactant.

The combustion chamber is tasked with making the required holding time available to the reactants for the duration of the reaction process at sufficiently high temperatures. The holding time in the thermal emission control systems is between 0.3 and 1.0 s. As combustion chambers are subject to high thermal stresses, scaling through exhaust gases containing O_2 and embrittlement caused by abrupt temperature changes, consideration must be first given to the correct material selection. In practice, all-metal combustion chambers and combustion chambers with metal shells and fire-resistant refractory lining have proven themselves. Combustion chambers that form a functional and inseparable unit with the burner—the so-called combustors (all-steel swirl combustion chambers)—have proven to be particularly favourable. Their benefit is that they have a small apparatus volume, endure high thermal combustion chamber stress, allow for simple and fast assembly and only require short initiation times due to the all-steel design.

The exhaust gas is tangentially directed into the combustor via an inlet flange in which it is moved by an adjustable swirl flap (1) in a swirling motion (4). This rotary flow, which ensures good reactant mixing, continues into the actual combustion chamber (5) via an external and internal annular gap (2). The internal cylinder is cooled through the annular gap flow-through with direction reversal. The exhaust air

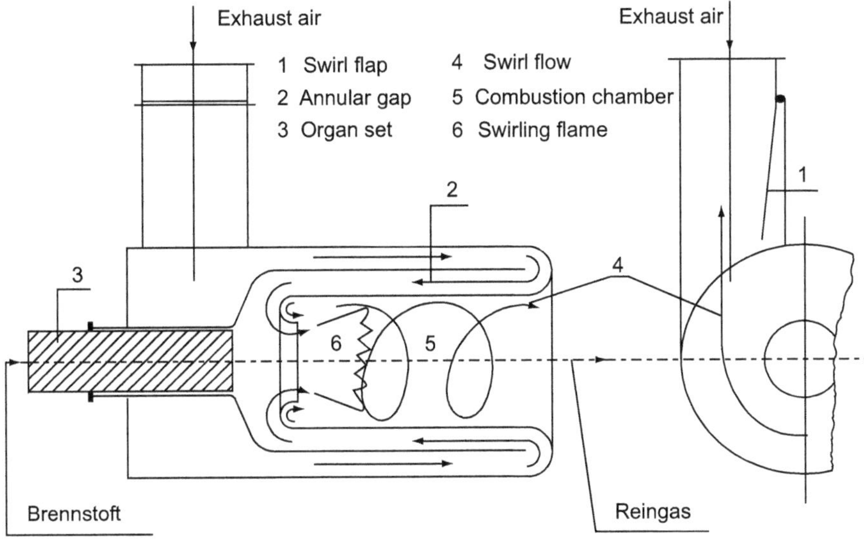

Fig. 8.10 Combustor. *1* Swirl flap, *2* annular gap, *3* organ set, *4* swirl flow, *5* combustion chamber, *6* swirling flame. (Company KEU (KEU 1989))

is then mixed with the combustion air and additional fuel, ignited and subsequently burned by the swirling flame (6) (Fig. 8.10).

One or more heat exchangers are placed downstream of the combustion chamber and/or combustor to achieve the highest possible efficiency in the TAR system. The heat exchangers are used on the one hand to heat the raw gas and on the other to preheat the combustion air and/or any necessary fresh air supply (omitted in the combustor procedure). This measure reduces energy expenditure, i.e. additional fuel consumption.

8.2.3 Combination Procedures

In addition to the individual procedures mentioned, there is also a series of combinations of these systems. Combination procedures are used if gaseous contaminants are present in addition to particulate contaminants, e.g. in power plants and waste incineration plants. These procedures work in the form of emission control systems in series.

8.3   Requirements for Emission Control Systems

In order to clean exhaust gases with the abovementioned emission control systems in a way that is effective, environmentally friendly and economical, said systems must fulfil a series of requirements (Kolar 1990). This includes:

- Adherence to the stipulated emission limit values and separation rates
- Environmentally friendly secondary processes

  - Minimal secondary emission
  - Absence of waste water
  - Adherence to noise limit values
  - Landscape protection

Recyclable residuals or residuals that can be dumped

- Low energy consumption

  - Minimal pressure loss
  - Reheating

- High availability

  - High state of development
  - Extensive operational experience
  - Low breakdown susceptibility

- Deadline compliance
- Low floor space requirement
- Low weight
- Low investments
- Minimal annual operating costs

In practice, not all requirements can be readily combined with one another; thus, it is necessary to set priorities based on the problem that needs to be solved.

8.4   Emission Control System Selection

Because primarily organic solvent mixtures in gaseous or vaporous aggregate states occur in coating companies, inertial separators, electrical separators and separators that filter and work when wet cannot be used. Their field of application lies—as described—in the removal of solid and liquid contaminants.

Thus, condensation, absorption, adsorption, sorption and oxidation remain as possible processes. Because the sorption processes are mainly applied to eliminate odorants, they—like condensation—are not applied, in particular since big disadvantages such as large construction volume, sensitivity to load variations and to toxic bacterial substances carry negative weight (VDI Guidelines 1977). However, condensation does not come into question for emission control as an individual process, as the required residual concentration can generally not be achieved with it.

In addition to organic solvents, raw gas in a coating company can also contain other contaminants, e.g. plasticizers. These generally cause damage to adsorbents when desorbed. Furthermore, the secondary removal of condensates, desorbates and/or washing solutions is extremely problematic with condensation, adsorption

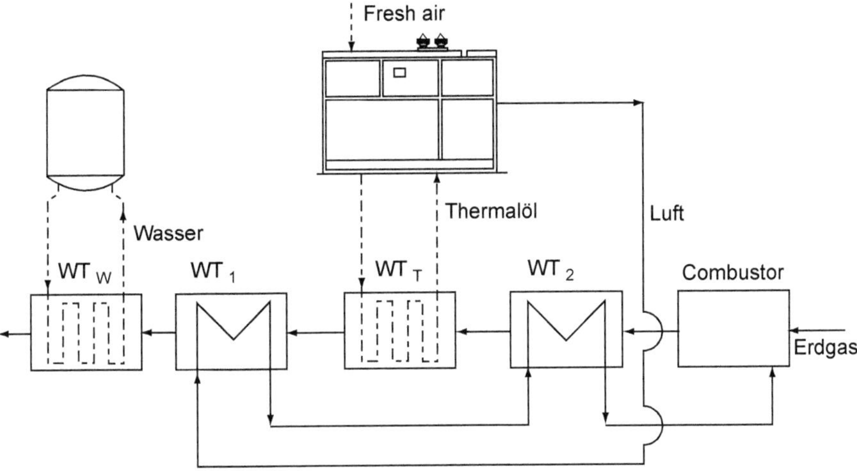

Fig. 8.11 TAR system with warm water supply

and absorption. However, the contaminants are converted into harmless components in oxidation processes.

Moreover, the limitation caused by using absorption is that chemical absorption can only be used for reactive substances and physical absorption can also only be used on specific substances. However, all organic substances—in particular solvents—are very inert. Another problem with adsorption is that organic vapours and/or organic compounds can make the adsorbent unusable through resinification.

According to the advantages and disadvantages listed, oxidation stands out as the most suitable process. In turn, thermal emission control is given preference, as it is more universally applicable than catalytic emission control because the catalytic converters can be damaged by contaminants (catalyst poisons). Particular attention is given to the fact that the catalytic converters in KAR systems become hazardous waste once their lifespan expires. Furthermore, the degree of cleaning in TAR systems is largely independent of the accrual of contaminants and thus remains high. For the reasons mentioned above, thermal emission control is the best alternative for coating companies.

Layout and Design of a TAR System For economical reasons, it is important to take heat recovery measures with a TAR system. Therefore, it is possible to preheat and heat exhaust air on the one hand and the required circulating thermal oil on the other. In practice it has been shown that it makes sense to install two exhaust air preheaters, whereby one is placed directly behind the combustion system and the other behind the thermal oil heat exchanger (Carlowitz 1984). The combustor has proven itself as a combustion system. As seen it Figs. 8.11 and 8.12, the exhaust air is heated by heat exchangers WT_1 and WT_2 and is then directed into the combustor. Out of the combustor comes the air that has been raised to combustion temperature, which transfers its energy in the form of heat to all three heat exchangers and cools itself

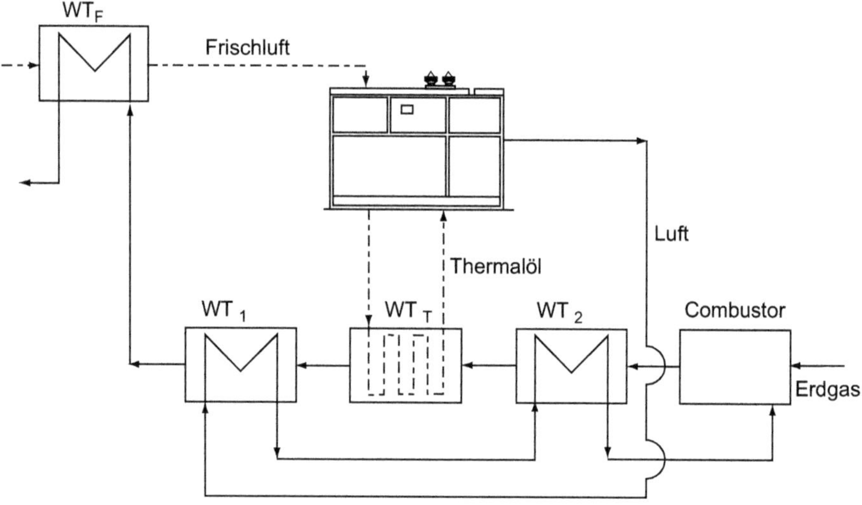

Fig. 8.12 TAR system with fresh air preheating

in the process. The thermal oil heat exchanger WT$_T$ reheats the return temperature
to the required flow temperature. Because the exhaust air still has a relatively high
energy level (100–200 °C), it can be used to guarantee the warm water supply,
for example, by placing a hot water boiler downstream. Even the fresh air can be
preheated, although this is generally not feasible since the often very high process
temperature cannot be achieved. Thus, one still needs an air preheater, particularly
since the expensive air ducts make the process economically unfeasible.

Thermal emission control is particularly suitable for the removal of halogen- and
chlorine-free contaminants in the textile industry. From an ecological point of view,
it is one of the most optimal solutions because energy is saved and contaminants are
completely removed at the same time (Table 8.3).

Azeotropic Mixtures

Table 8.3 Standard solvents and their characteristics

	Boiling point (°C)		Boiling point (°C)	Composition boiling point in (°C)	Shared in weight %
Acetone	56	Amyl acetate	140	97.5:2.5	131.3
Acetone	256	Methanol	65	14:86	55.7
Amyl acetate	140	Acetone	56	2.5:97.5	131.3
Benzene	80	Ethanol	78.3	67.6:32.4	68.2
Benzene	80	Methanol	65	60:40	58.3
Benzene	80	w propyl alcohol	95–97	83:17	77.1
Butanol	114–120	Butyl acetate	121–127	47:53	117.2
Butanol	114–120	Toluene	111	32:68	105.5

Table 8.3 (continued)

	Boiling point (°C)		Boiling point (°C)	Composition boiling point in (°C)	Shared in weight %
Butanol	114–120	Water	100	63:37	92.3
Butyl acetate	121–127	Butanol	114–120	53:47	117.2
Butyl acetate	121–127	n propyl alcohol	95–97	60:40	94.2
Ethanol	78.3	Benzene	80	32.4:67.6	68.2
Ethanol	78.3	Ethyl acetate	74–77	30.6:69.4	71.8
Ethanol	78.3	Carbon tetrachloride	77–78	16:84	64.9
Ethanol	78.3	Toluene	111	68:32	76.7
Ethanol	78.3	Water	100	95.6:4.4	78.1
Ethyl acetate	74–77	Ethanol	78.3	69.4:30.6	71.8
Ethyl acetate	74–77	Methanol	65	81:19	54
Ethyl acetate	74–77	Carbon tetrachloride	77–78	43:57	74.8
Isopropyl alcohol	80–82	Methyl acetate	56–62	70:30	77.3
Methanol	65	Acetone	56	86:14	55.7
Methanol	65	Ethyl acetate	74–77	74:77	54
Methanol	65	Benzene	80	40:60	58.3
Methanol	65	Methyl acetate	56–62	19:81	54
Methyl acetate	56–62	Isopropyl alcohol	80–82	30:70	77.3
Methyl acetate	56–62	Methanol	65	81:19	54
n propyl alcohol	95–97	Benzene	80	17:83	77.1
n propyl alcohol	95–97	Butyl acetate	121–127	40:60	94.2
Carbon tetrachloride	77–78	Ethyl acetate	74–77	57:43	74.8
Carbon tetrachloride	77–78	Ethanol	78.3	84:16	64.9
Carbon tetrachloride	77–78	Methanol	65	79:21	55.7
Toluene	111	Ethanol	78.3	32:68	76.7
Toluene	111	Butanol	114–120	68:32	105.5
Water	100	Ethanol	78.3	4.4:95.6	78.1

Chapter 9
Investment Profitability Analysis

In order to make final decisions regarding investment and/or on the selection from two or more alternatives, a profitability analysis focused specifically focused on the coating industry is provided below. Not just costs, but also company strategies are taken into consideration.

Decisions on investments influence a company's future success or failure. These must therefore be carefully planned and executed if the desired corporate goals are to be achieved.

Figure 9.1 shows the most important investment goals.

The main objective of an investment in textile coating is to have the right selection of investment goods and to purchase those goods which can produce assets through the transformation of production factors. An investment in tangible assets aims to identify and acquire the investment object which promises the most favourable requirements for the rational and profitable design of a production process in textile finishing.

The traditional investment is often no longer enough to substantiate the profitability of investments in new production technologies in an optimum manner. The classical procedure considers an investment from a purely monetary perspective only and neglects the integration effect of performance and costs. Moreover, it must shoulder the blame for the exclusion of strategic investment effects and lack of integration in cost calculation (Singer 1991). It is therefore advisable to abandon the previous purely investment approach and to migrate to profitability with consideration given to corporate and product planning.

All investments go through a decision process. Said process can be executed as depicted in Fig. 9.2 with regard to profitability analysis in the above mentioned terms and to basic tasks.

In order to assess the profitability of an investment, the output—the production performance—and the input, i.e. the production costs, must be taken into consideration. Output in the coating industry is the quantity in square metres which must be produced per year, the organisation to be coordinated with the plant, the ability to have a variety of products, the necessary changeover times and the ability to produce various levels of quality. In addition to pure production costs, input consists of the costs for the use of space—warehouse costs—as well as for factory planning.

A. Giessmann, *Coating Substrates and Textiles,*
DOI 10.1007/978-3-642-29160-9_9, © Springer-Verlag Berlin Heidelberg 2012

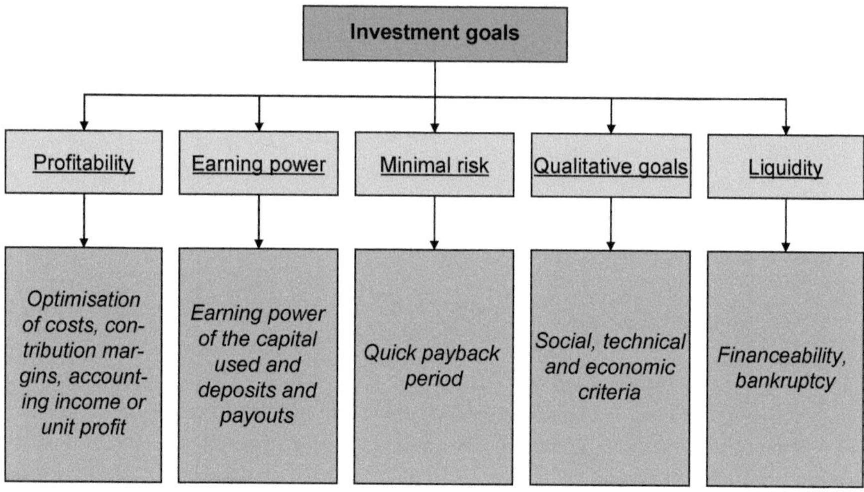

Fig. 9.1 Investment goals. (REFA 1985)

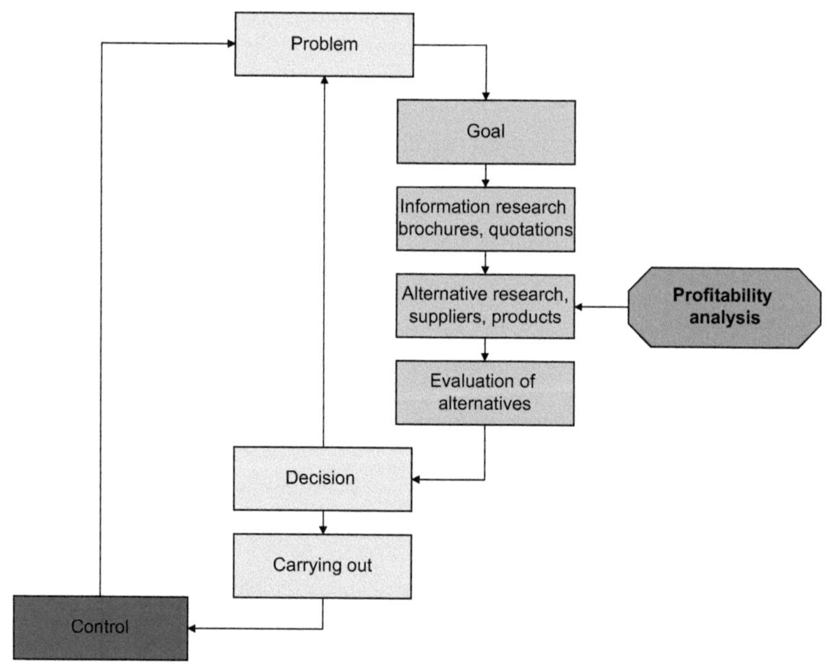

Fig. 9.2 Decision process. (Pupka 1983)

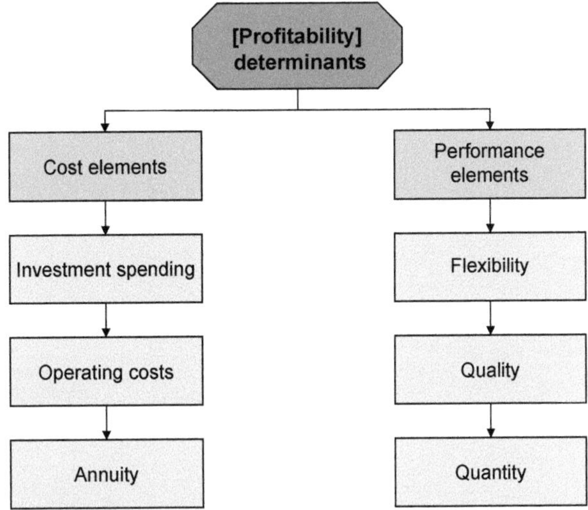

Fig. 9.3 Profitability determinants. (Singer 1991)

Therefore, in the example for the purchase and operation of a coating plant, investment is not just observed in terms of cost factors, but also in terms of performance elements (Fig. 9.3).

Investment objects with a high level of automation are only profitable if considerable production performance is required. Conventional factories are far more profitable if only a low machine production is required to meet demand due to the market situation. Production technologies with the highest possible performance ratio only make sense if the goods produced can also be sold. Profitability can be calculated using the following determinants:

$$W = \frac{LF + KF}{2}$$

W = Profitability
LF = Performance elements
KF = Cost elements

The result of a profitability analysis is a concrete recommendation for action. This means, for example, a statement on which of two investment objects should be purchased because it promises higher profitability.

9.1   Strategic Profitability Analysis

The strategic profitability analysis links the investment in the purchase of a textile finishing machine with strategic corporate and product planning. Its advantages lie in the fact that uniform standards apply in the entire production area and so-called isolated applications are avoided.

Fig. 9.4 Investment development

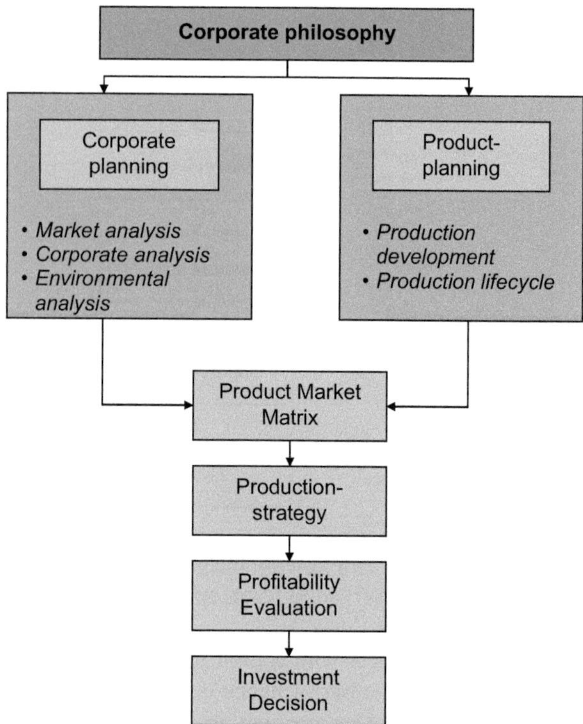

Figure 9.4 shows the connections that must be taken into consideration when making an investment.

The basic goals, moral concepts and intentions of the company are identified as the corporate philosophy.

9.2 The Business Strategy

The tasks of strategic corporate planning—long-term planning for 5–10 years—is the systematic development of operational corporate goals and of measures to realise them.

Corporate planning uses various analyses as a requirement for strategic formulation. A company's three most important basic strategies result from application and assessment of said analyses (Fig. 9.5).

In this way, market leadership for a coating company in standard products, i.e. Imitation leather in various designs and with various characteristics, would be desirable for the shoe, belt and pillow industry.

Differentiation means that the producer provides exceptional, services beyond the usual in specific product areas, e.g. delivery in the shortest time or direct from stock.

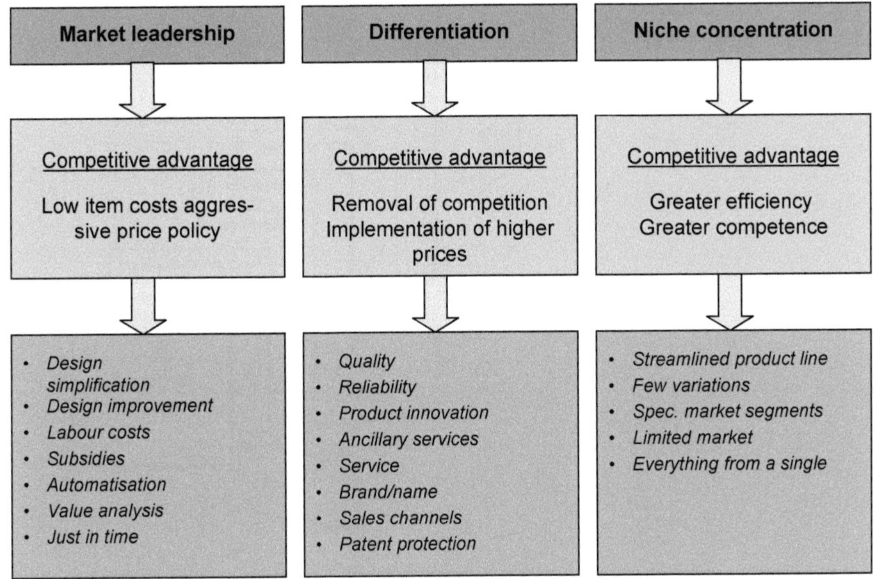

Fig. 9.5 Basic strategies for corporate planning. (Appell 1981)

The production of technical textiles is an example of a niche concentration. The technically sophisticated ware can only be made by a few companies due to the elaborate production process and the required specialist knowledge.

9.3 Production Strategy

In addition to product quality and the cost structure, the time factor continues to gain importance. The company with the shortest lead-time from the product idea to the deliverable product gains decisive competitive advantages. Product planning is understood as a uniform, market-oriented way of thinking which accompanies the product from concept to technical development and launch. It includes all influences, decisions and actions that are temporally and technically recorded, verified and managed (Hinterhuber 1966).

Three product planning components are derived from the constant interaction of product, market and company and taking into account high innovation rates (Geyer 1987).

Methodical Coordination This involves coordination of technical, ecological and economical information before releasing a product for product development. The analytical development of ideas and idea evaluation are criteria for methodical decision-making preparations.

Monitoring Constant monitoring of a product's lifecycle must be guaranteed.

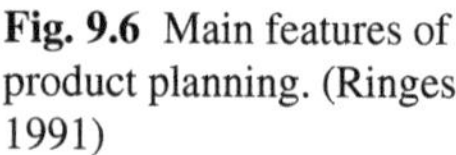

Fig. 9.6 Main features of product planning. (Ringes 1991)

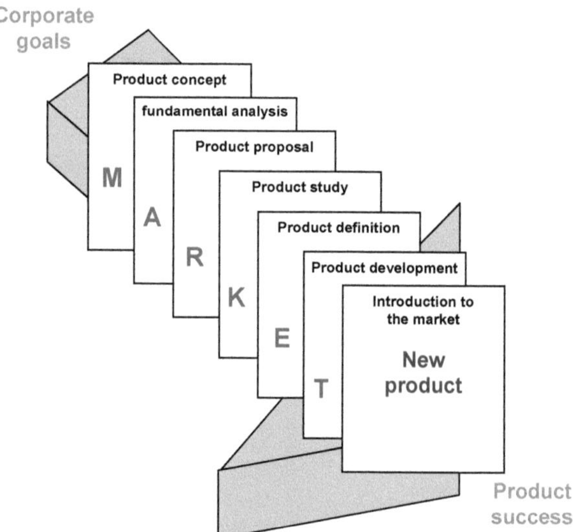

Leadership Rules and Code of Conduct Work in the team and the interaction between corporate management can be optimised via suitable leadership guidelines.

Product planning begins with the specification of corporate goals and leads to the steps shown and finally to product success (Fig. 9.6).

Concrete actions can be taken using the knowledge gained from corporate and product planning. The product market matrix helps make a decision regarding which products can lead to success on which markets.

Experience has shown that, for most companies, investment in a coating plant represents a horizontal diversification. The company is thereby required to make fundamental decisions with regard to the production programme (Fig. 9.7).

The current world market situation shows that, in particular, imitation leather, vinyl floors and clothing materials have a large market share. One could then conclude that a coating company has ideal distribution for production:

- approx. 70 % imitation leather,
- approx. 15 % vinyl floor,
- approx. 15 % clothing materials.

9.4   Coating Plant Cost Elements

The cost elements are divided into a series of expenses incurred for an investment in a coating company.

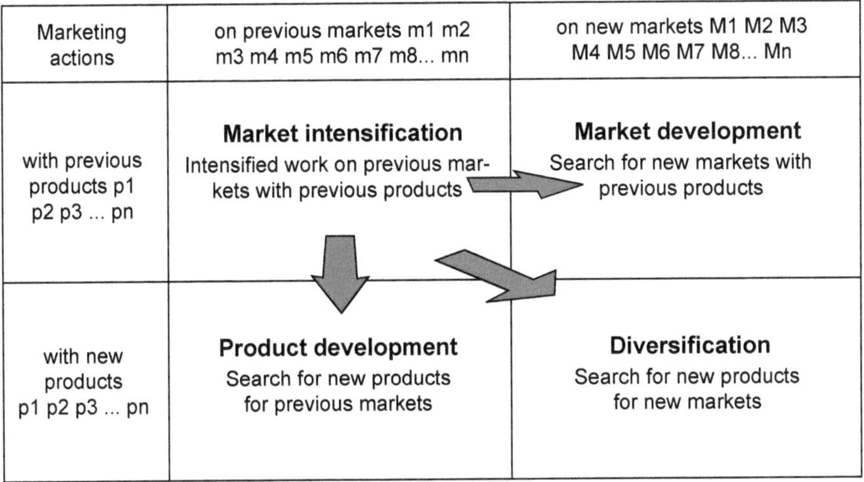

Fig. 9.7 Product market matrix. (Galweiler 1989)

Table 9.1 Total investment

Total costs (in €)	Tridem	Tandem
Machines (SKA)	4,120,400-	3,483,000
Additional units (SKZ)	392,500-	325,000
Development (SKP)	1,063,000-	980,000
Freight and transport (SKT)	508,875-	372,500
Factory costs (SKF)	1,900,000-	1,900,000
Total SEK	7,984,775	7,060,500

9.4.1   Investment Spending

All expenses incurred for a coating company are considered investment spending. The prices, costs and values specified are purely fictional and in no way represent reality. The example in Table 9.1 shows the comparison between two-coat machines (Tandem) and three-coat machines (Tridem).

This would include machines, development, the factory building including the entire inventory and additional units.

9.4.2   Operating Costs

Operating costs (SBT) include all costs incurred in the coating company during a year period. This includes personnel costs (SPK), energy costs (SEK), maintenance costs (SWK), insurance costs (SVK) and amortisation costs (SAK). The annual operating times—300 workdays—are generally placed at 5,400 h in two-shift operation and

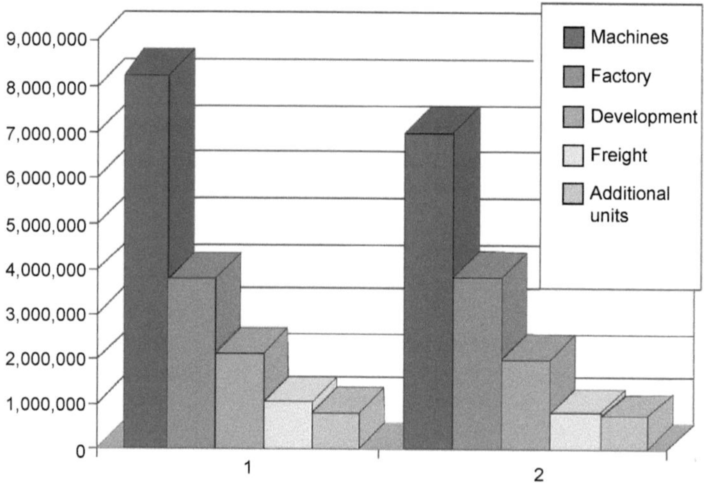

Fig. 9.8 Graphic comparison of total investments

2,900 h in one-shift operation (Fig. 9.8).

$$SBT = SPK + SEK + SWK + SVK + SAK$$

9.4.2.1 Personnel Costs

Personnel expenses for both machines is the same since the higher automation of the Three-coat machine covers the additional personnel expenses usually required when compared to the two-coat machine.

The following personnel costs result when consideration is given to the salary structure of developing countries and emerging markets:

a. One-shift operation with 2,900 h/a: SPK   196,000 €/a
b. Two-shift operation with 5,400 h/a: SPK   267,200 €/a.

9.4.2.2 Energy Costs

The accrued energy costs are comprised of: electricity costs (KES), water costs (KEW), compressed air costs (KED) and oil costs (KEÖ).

Three-coat machine (Tables 9.2 and 9.3)

Two-coat machine (Tables 9.4 and 9.5).

Table 9.2 One-shift operation with 2,900 h/a

Energy	Consumption (h^{-1})	Price (€/E)	Consumption (a^{-1})	Total (€/a)
Electricity (KVA)	800/60 %	0.03	1,392,000	41,760
Oil (kg)	200	0.13	580,000	7,540
Water (m^3)	12	0.15	34,800	5,220
Compressed air (bar)	10	0.05	29,000	1,450
Total (SEK)				123,830

Table 9.3 Two-shift operation with 5,400 h/a

Energy	Consumption (h^{-1})	Price (€/E)	Consumption (a^{-1})	Total (€/a)
Electricity (KVA)	800/60 %	0.03	2,592,000	77,760
Oil (kg)	200	0.13	1,080,000	140,400
Water (m^3)	12	0.15	64,800	9,720
Compressed air (bar)	10	0.05	29,000	1,450
Total SEK				229,330

Table 9.4 One-shift operation with 2,900 h/a

Energy	Consumption (h^{-1})	Price (€/E)	Consumption (a^{-1})	Total (€/a)
Electricity (KVA)	600/60 %	0.03	1,044,000	31,320
Oil (kg)	150	0.13	435,000	56,550
Water (m^3)	8	0.15	23,200	3,480
Compressed air (bar)	10	0.05	29,000	1,450
Total SEK				92,800

Table 9.5 Two-shift operation with 5,400 h/a

Energy	Consumption (h^{-1})	Price (€/E)	Consumption (a^{-1})	Total (€/a)
Electricity (KVA)	600/60 %	0.03	1,944,000	56,320
Oil (kg)	150	0.13	810,000	105,300
Water (m^3)	8	0.15	43,200	6,480
Compressed air (bar)	10	0.05	29,000	1,450
Total SEK				171,550

9.4.2.3   Maintenance and Repair Costs

Reference values for the maintenance and repair costs of the KWA plants and the KWG building are set at 1 and 0.5 % of the acquisition value, respectively. For two-shift operation the value is set at 1.2 %.

$$KWA1 = SKA \times 0.01/a$$

$$KWA2 = SKA \times 0.012/a$$

$$KWG = SKF \times 0.005/a$$

$$KW_{ges} = KWA(1.2) + KWG$$

This results in the following costs:

$$KWG = 1{,}900{,}000 \, € \times 0.005/a$$
$$= 9{,}500 \, €/a.$$

Three-Coat Machine

a. One-shift operation with 2,900 h/a

$$KWA1 = 4{,}120{,}400 \, € \times 0.01/a$$
$$= 41{,}204 \, €/a$$
$$SWK = 41{,}204 \, €/a + 9{,}500 \, €/a$$
$$= 50{,}704 \, €/a$$

b. Two-shift operation with 5,400 h/a

$$KWA2 = 4{,}120{,}400 \, € \times 0.012/a$$
$$SWK = 49{,}445 \, €/a + 9{,}500 \, €/a$$
$$= 58{,}945 \, €/a$$

Two-Coat Machine

a. One-shift operation with 2,900 h/a

$$KWA1 = 3{,}483{,}000 \, € \times 0.01/a$$
$$= 34{,}830 \, €/a$$
$$SWK = 34{,}830 \, €/a + 9{,}500 \, €/a$$
$$= 44{,}330 \, €/a$$

b. Two-shift operation with 5,400 h/a

$$KWA2 = 3{,}483{,}000 \, € \times 0.012/a$$
$$= 41{,}796 \, €/a.$$
$$SWK = 41{,}796 \, €/a + 9{,}500 \, €/a$$
$$= 51{,}296 \, €/a$$

9.4.2.4   Other Costs

Insurance premiums for the building and for the plant fall under other costs. The
insurance costs can also be determined using reference values:

$$KVA = SKA \times 0.015/a$$
$$KVG = SKF \times 0.008/a.$$

Which results in:

$$KVG = 1{,}900{,}000 \, € \times 0.008/a$$
$$= 15{,}200 \, €/a.$$

Three-Coat Machine

$$KVA = 4{,}120{,}400 \, € \times 0.015/a$$
$$= 61{,}800 \, €/a$$

Two-Coat Machine

$$KVA = 3{,}483{,}000 \, € \times 0.015/a$$
$$= 52{,}245 \, €/a$$

The total costs for insurance are calculated with:

$$SVK = KVA + KVG.$$

Three-Coat Machine

$$SVK = 61{,}800 \, €/a + 15{,}200 \, €/a$$
$$= 77{,}006 \, €/a$$

Two-Coat Machine

$$SVK = 52{,}245 \, €/a + 15{,}200 \, €/a$$
$$= 67{,}445 \, €/a$$

9.4.2.5   Amortisation

Amortisation is linear in all cases. Each year has the same weight, with the exception that a building has an operating life of 20 years and the machines and their additional units have an operating life of 10 years. The following formulas apply:

$$KAG = SKF/20 \, a,$$
$$KAA = SKA/10 \, a,$$
$$KAZ = SKZ/10 \, a.$$

The formulas have to be changed from 2,900 to 5,400 h/a for two-shift operation, which corresponds to a factor of 1.86.

Three-Coat Machine

One-shift operation	Two-shift operation
Building	
KAG = 1,900,000 €/20a	95,000 €/a × 1.86
95,000 €/a	176,700 €/a
Machine	
KAA = 4,120,400 €/10a	412,040 €/a × 1.86
412,040 €/a	766,394 €/a
Additional units	
KAZ = 392,500 €/10a	39,250 €/a × 1.86
39,250 €/a	73,005 €/a
Total *SAK = 546,290 €/a*	*1,016,099 €/a*

Two-Coat Machine

One-shift operation	Two-shift operation
Building	
KAG = 1,900,000 €/20a	95,000 €/a × 1.86
95,000 €/a	176,700 €/a
Machine	
KAA = 3,483,000 €/10a	348,300 €/a × 1.86
348,300 €/a	647,838/a
Additional units	
KAZ = 325,000 €/10a	32,500 €/a × 1.86
32,500 €/a	60,450 €/a
Total *SAK = 475,800 €/a*	*884,988 €/a*

9.4.3 *Annuity*

Annuity reflects the annual weight for repayment and interest. To develop a realistic image, it is assumed that 75,000 € can be raised as equity so that the flat values can be further calculated:

$$7,984,775\ € - 734,775\ € = 7,250,000\ €\ \text{or}$$

$$7,060,500\ € - 710,500\ € = 6,350,000\ €$$

This would mean the following, assuming that a bank grants credit with a maturity of 10 years with 8 % interest:

$$A = IK \times a$$

$$P \times (1 + P)$$

Table 9.6 Cost comparison between two-coat machine and three-coat machine

	Plant 1		Plant 2	
	1st shift	2nd shift	1st shift	2nd shift
Personnel costs	196,000	267,200	196,000	267,200
Maintenance and repair costs	50,704	58,945	44,300	51,296
Energy costs	123,830	229,330	92,800	171,550
Other costs	77,006	77,006	67,445	67,445
Amortisation	546,290	1,016,099	475,800	884,988
Total SBT	993,830	1,648,580	876,345	1,442,479
Annuity	1,080,250	1,080,250	946,150	946,150
Total SKJ	2,074,080	2,728,830	1,822,495	2,388,629

In the calculation example, a is:

$$-\frac{0.08 - (1 + 0.08)^{10}}{(1 + 0.08)^{10} - 1}$$

$$= 149 \Rightarrow 14.9\,\%.$$

The annuities for both plants can be calculated:

Three-coat machine

$$A = 7{,}250{,}000\,\text{€} \times 0.149$$

$$= 1{,}080{,}250\,\text{€/a}$$

Two-coat machine

$$A = 6{,}350{,}000\,\text{€} \times 0.149$$

$$= 946{,}150\,\text{€/a}$$

9.4.4 Total Costs

Table 9.6 shows that it always makes sense to operate in two-shift operation. Almost twice as much can be produced, whereby the costs increase 30 % in proportion. Thus, two-shift operation is assumed in the following sections.

9.5 Performance Elements in Coating Plants

Like the cost elements, the performance elements must be more closely examined. These elements include flexibility, quality and quantity.

Table 9.7 Profitability of a three-coat machine

	Days (d/a)	Time (min/a)	v (m/min)	b (m)	n (%)	Amount (m²/a)
Imitation leather	200	1,080	15	1.42	87	4,002,696
Vinyl floor	50	1,080	6	2.00	87	563,760
Clothing materials	50	1,080	18	1.60	87	1,353,024
Total	300				87	5,919,480

Table 9.8 Profitability of a two-coat machine

	Days (min/a)	Time (m/min)	v (d/a)	b (m)	n (%)	Amount (m²/a)
Imitation leather	130	1,080	12	1.42	87	2,081,402
	70		7	1.42		653,774
Vinyl floor	50	1,080	5	2.00	87	469,800
Clothing materials	50	1,080	15	1.60	87	1,127,520
Total	300				87	4,332,496

9.5.1   Flexibility

All materials described are produced according to various coating methods. Production is performed in the so-called single-coat, tandem or tridem coating system. This means one, two or three coating applications are required with the corresponding drying and setting processes. In the example it is assumed that both machines can be adjusted at any time to fluctuating market situations—caused by fashion changes and technical requirements. The three-coat machine has an advantage over the two-coat machine in terms of flexibility thanks to its application systems. Reintroduction for additional coating processes is omitted.

9.5.2   Quality

All product qualities found on the market should be able to be produced with both machines. For the machines, the quality obtained depends mainly on the formulas and thus on the operator's experience and know-how.

9.5.3   Quality

The three-coat machine has an output of 15 m/min when producing imitation leather. In the tandem system, capacity must be assumed to be 12 m/min for 67 % of the capacity and 7 m/min for the remaining 23 % when making imitation leather, which requires a triple coat. These speeds result from the fact that one can directly run the first two treatment processes and that the third process necessitates the re-feeding of the machine. All other speeds can be found in Tables 9.7 and 9.8. Under the

Table 9.9 Determining weighting factors

	IK	BK	JK	FB	QL	QN	SM	GF
Investment costs		2	1	0	0	1	4	0.14
Operating costs	0		1	0	1	1	3	0.10
Annual GK	1	1		1	2	2	7	0.23
Flexibility	2	2	1		1	1	7	0.23
Quality	2	1	0	1		2	6	0.20
Quantity	1	1	0	1	0		3	0.10
Total	6	7	3	3	4	7	30	1.00

assumption that of the 300 available workdays, 200 days are used for the production of imitation leather, 50 days are used for the production of vinyl floors and 50 days are used for the production of clothing materials, this means that for the tandem system, 200 days must be divided into 130 days with 12 m/min and 70 days with 7 m/min. Experience dictates that the capacity utilisation in textile finishing plants is between 85 and 90 %. The value of 87 % is assumed as the average in the following calculations. 18 h/day, or 1,080 min, are assumed for two-shift operation.

9.6 Operational Profitability Analysis

The operational profitability analysis compares the cost and performance element requirements derived from the business and product strategies with the degree and risk of performance expected in the product strategy. For this, the weighting and compliance factors of the individual cost and performance elements are required.

All factors are rated to obtain a suitable weighting factor (GF). There are three options:

$2{:}0 \rightarrow \bullet$ 1st factor is more important than the 2nd factor,
$1{:}1 \rightarrow \bullet$ Both factors are equally important.
$0{:}2 \rightarrow$ 2nd factor is more important than the 1st factor.

The weighting factors subjectively determined in Table 9.9 apply to both machines.

In order to determine the compliance factors (EF), it is necessary to have a conversion formula and a table that show which compliance factor is present at a specific grading (for factors that are not numerically ascertainable).

Here, the following equations are used:
Case 1: Factors with the same classification

$$EF = 1 + \frac{9 \times (FA - FA_{min})}{FA_{max} - FA_{min}}$$

Case 2: Factors with opposing classification

$$EF = 1 + \frac{9 \times (FA_{max} - FA)}{FA_{max} - FA_{min}}$$

Table 9.10 Determining the compliance factor from the assessed factors

FA assessment	EF
Sufficient	2.10
Satisfactory/sufficient	3.30
Satisfactory	4.30
Good/satisfactory	5.50
Good	6.60
Very good/good	7.80
Very good	8.90

EF = compliance factor

FA = factors

Note: Whether a classification is the same or opposed depends on whether the higher numerical value is better (case 1) or the converse (case 2).

The compliance factor can be taken from Table 9.10 for the factors ranked.

The maximum and minimum values for the factors are determined based on the company's financial situation and business and product strategies.

Three different qualities must be calculated for the three different products—based on the various speeds.

Profitability can then be determined using the data identified. The probability factor included covers any deviations that may occur under certain circumstances. Because the performance elements from the profitability analysis are different for the three products due to their various quantities and desirabilities (FA_{min}, FA_{max}), they are shown separately (while the cost elements are not).

The key profitability figures W can be determined using the calculated values according to the specified basic formula (Tables 9.11, 9.12, 9.13, 9.14, 9.15, 9.16, 9.17 and 9.18).

Table 9.11 Calculating the compliance factor EF

	Tridem	EF	Tandem	EF
$FA_{max} = 9{,}000{,}000$ Investment costs [D] $FA_{min} = 5{,}000{,}000$	7,984,775	3.28	7,060,500	5.36
$FA_{max} = 2{,}500{,}000$ Operating costs [D] $FA_{min} = 500{,}000$	1,649,580	6.17	1,442,419	5.24
$FA_{max} = 4{,}000{,}000$ Annual GK [D] $FA_{min} = 1{,}500{,}000$	2,728,830	4.86	2,388,629	6.18
$FA_{min} =$ sufficient Flexibility [/] $FA_{max} =$ very good	Very good	8.90	Good/satis.	5.50
$FA_{min} =$ sufficient Quality [/] $FA_{max} =$ very good	Very good	8.90	Very good	8.90

Table 9.12 Imitation leather

$FA_{min} = 2,000,000$ Quantity (m^2) $FA_{max} = 5,000,000$	4,002,696	7.00	2,735,176	3.21

Table 9.13 Vinyl floors

$FA_{min} = 200,000$ Quantity (m^2) $FA_{max} = 1,000,000$	563,760	5.09	469,800	4.04

Table 9.14 Clothing materials

$FA_{min} = 750,000$ Quantity (m^2) FA_{max}	1,353,024	6.43	1,127,520	4.40

Table 9.15 Cost element profitability factors

Cost elements Machine	GF	Compliance factor (EF)		EF risk	Value	
		Tridem	Tandem		Tridem	Tandem
Investment costs	0.14	3.28	5.36	100	0.46	0.75
Operating costs	0.10	6.17	5.24	100	0.62	0.52
Annual GK	0.23	4.86	6.18	105	1.17	1.49
Total					2.25	2.76

Table 9.16 Imitation leather

Cost elements Machine	GF	Compliance factor (EF)		EF risk	Value	
		Tridem	Tandem		Tridem	Tandem
Flexibility	0.23	8.90	5.50	110	2.25	1.39
Quality	0.20	8.90	8.90	100	1.78	1.78
Quantity	0.10	7.00	3.21	120	0.84	0.39
Total					4.78	3.56

Table 9.17 Vinyl floors

Cost elements Machine	GF	Compliance factor (EF)		EF risk	Value	
		Tridem	Tandem		Tridem	Tandem
Flexibility	0.23	8.90	5.50	110	2.25	1.39
Quality	0.20	8.90	8.90	100	1.78	1.78
Quantity	0.10	5.09	4.04	120	0.61	0.48
Total					4.64	3.65

Table 9.18 Clothing materials

| Cost elements | GF | Compliance factor (EF) | | EF risk | Value | |
Machine		Tridem	Tandem		Tridem	Tandem
Flexibility	0.23	8.90	5.50	110	2.25	1.39
Quality	0.20	8.90	8.90	100	1.78	1.78
Quantity	0.10	6.43	4.40	120	0.77	0.53
Total					4.80	3.70

Table 9.19 Determining the key profitability figures W

	Tridem	W	Tandem	W
Imitation leather	(4.78 + 2.30)/2	3.54	(3.56 + 3.01)72	3.29
Vinyl floors	(4.64 + 2.30)/2	3.47	(3.65 + 3.01)72	3.33
Clothing materials	(4.80 + 2.30)72	3.55	(3.70 + 3.01)72	3.36

Table 9.20 Annual sales for production with a three-coat machine

	Amount (m^2/a)	Price ($€/m^2$)	Sales ($€$/a)
Imitation leather	4,002,696	2.28–	9,126,146–
Vinyl floor	563,760	3.75–	2,114,100–
Clothing materials	1,353,024	1.60–	2,164,834–
Total	5,919,480		13,405,080–

As shown in Table 9.19, the three-coat machine achieves a higher value for all three products, which shows that the performance and cost elements guarantee high profitability with regard to the business and product strategies to be followed.

For most companies, the amortisation period for their investments plays an important role. Thus, a connection between amortisation and the calculated key profitability figures is established below using an annual overview.

9.7   Planning Result Calculation

Specific profits are aimed for when a coating company is established. An annual overview, which should provide an insight into the anticipated revenues and profits to be expected, follows below.

9.7.1   Annual Sales

The following example sales prices are assumed for the calculation of annual sales (Tables 9.20 and 9.21):

Imitation leather	2.28 $€/m^2$
Vinyl floor	3.75 $€/m^2$
Clothing materials	1.60 $€/m^2$

Table 9.21 Annual sales for production with a two-coat machine

	Amount (m^2/a)	Price ($€$/m^2)	Sales ($€$/a)
Imitation leather	2,735,176	2.28–	6,236,201–
Vinyl floors	469,800	3.75–	1,761,750–
Clothing materials	1,127,520	1.60–	1,804,032–
Total	4,332,496		9,801,983–

Table 9.22 Material costs for a three-coat machine

	Amount (m^2/a)	Price ($€$/m^2)	MK ($€$/a)
Imitation leather	4,002,696	1.28	5,123,450–
Vinyl floor	563,760	1.22–	687,787–
Clothing materials	1,353,024	0.65–	879,465–
Total	5,919,480		6,690,702–

Table 9.23 Material costs for a two-coat machine

	Amount (m^2/a)	Price ($€$/m^2)	MK ($€$/a)
Imitation leather	2,735,176	1.28	3,501,025
Vinyl floor	469,800	1.22	573,156
Clothing materials	1,127,520	0.65	732,888
Total	4,332,496		4,807,069

The expected sales (U) is based on multiplication of the amount sold (m) with the sales price achieved (K):

$$U = m \times K.$$

9.7.2   Profits

The annual total costs (JK) and the material costs (MK) must be subtracted from the target sales in order to calculate the possible annual profits (G) (Braune 1981):

$$G = U - (JK + MK).$$

Tables 9.22 and 9.23 show fictional material costs for the three products, i.e. imitation leather, vinyl floors and clothing materials. On the one hand, they are based on the individual prices of the respective suppliers; on the other, they are based on the formulas used.

9.7.3   Summary of All Costs and Profits

Table 9.24 provides an overview of all incomes and expenses for two-coat machines and three-coat machines.

Table 9.24 Comparison of all costs of two-coat machines and three-coat machines

	Tridem	Tandem
Sales (€/a)	13,405,080−	9,801,983−
Total costs (€/a)	2,728,830	2,388,629
Material costs (€/a)	6,690,702−	4,807,069−
Profits (€/a)	3,985,548−	2,606,285−
Amortisation (€/a)	1,016,099	884,988
Return flow (€/a)	5,001,647−	3,491,273−
Amortisation period (a)	1.60	2.02

Table 9.25 Comparison of all costs of two-coat machines and three-coat machines

	Tridem	Tandem
Imitation leather	2.25	2.44
Vinyl floor	2.22	2.46
Tent and rain materials	2.25	2.47
Average 0	2.24	2.46

The amortisation period is calculated by dividing the investment costs (IK) by the return flow (R) (Olfert 1988):

$$A = \frac{IK}{R}.$$

In practice, an amortisation period of less than 3 years is required due to reasons of risk. This means that all investments with longer amortisation periods are omitted from acquisition consideration from the start (Wohe 1990). Both plants described are thus clearly below the industry required amortisation period.

9.7.4 Comparison Figures

Every company desires and expects a short amortisation period. The calculated amortisation periods can be put in a 2:1 ratio with the key profitability figures, thereby calculating new comparison figures (Table 9.25)

$$(VKZ) = \frac{(VKZ = (2 \times A) + W)}{3}.$$

The lower the comparison figures, the more profitable the investments. The comparison figures, which are based on the individual products, are obtained using the abovementioned equation.

The average comparison figure is about as high for both plants. However, the three-coat machine proves itself to be the more profitable variation with consideration given to the amortisation period A and the key profitability figures W.

The example shows how complex a consideration of various plant layouts is and the methods via which a profitability study can be completed.

Bibliography

1. AKB Firmenprospekt (1994): Prozeßwärme mit AKB-Systemen. Sinsheim 1994
2. Appelt, H. (1981): Die neuen Techniken der Unternehmensplanung. Munich: Oldenbourg 1981
3. Arbeitgeberkreis Gesamttextil (Hrsg.) (1992): Textilveredlung Beschichten. Ausbildungsmittel-Unterrichtshilfen. Eschborn 1992
4. Baumbach, G. (1990): Luftreinhaltung. Berlin, Heidelberg, New York: Springer 1990
5. Berndt, H.-J. (1990): PVC Schaumkunstleder vom Stahlband. Sonderdruck Textilbericht 1990, S. 15–19
6. BG Chemie (Hrsg.) (1989): Richtlinien für die Vermeidung von Zündgefahren. Richtlinie Nr. 4. 1989
7. Birkle, M. (1979): Meßtechnik für den Immissionsschutz. Munich: Oldenbourg 1979
8. Boisseree, K. u. a. (1962): Immissionsschutzrecht in NRW. Siegburg: Reckinger ampl Co. 1962
9. Brathuhn, R. (1986): Pastenverarbeitung und Pulverbeschichtung. Munich: Hanser 1986
10. Braune, G. (1981): Betriebswirtschaftslehre für technische Fachkräfte. Bd. 37. Grafenau: Expert 1981
11. Breuer, H. (1987): dtv-Atlas zur Chemie. Bd. 1. Munich: Deutscher Taschen buch Verlag 1987
12. Bundesimmissionsschutzgesetz BImschG idFv. 14. May 1990. § 3.1
13. Bundesimmissionsschutzverordnung BImschV idFv. 15. July 1988. § 2.4
14. Carlowitz, O. (1984): Thermische Abluftreinigung. Krefeld: Technologie Report 1984
15. CG-Tee Homepage: Vorteile von CFK http://www.cg-tec.de/Vorteile_von_CFK.93.0.html (Letzter Zugriff 21.12.08)
16. Coatema Firmenprospekt (1990): Maschine für die Textilindustrie. Neuss: 1990
17. Cohen, E.G. (1992): Modern Coating and Drying Technology, Wiley-VCH 1992
18. Delignit. http://www.delignit.de/contenido/cms/upload/pdf/presse/Wissenswertes_ueber ESH-Lack.pdf (Letzter Zugriff 19.01.09)
19. Dornbusch Firmenprospekt (1994): Walzentechnologie. Krefeld 1994
20. Dufke, D. (1989): Herstellung von Leder. Leipzig: VEB Fachbuchverlag 1989
21. Echtermeyer, H. (1986): Veredlung von Textilien. 3. Aufl. Leipzig: VEB Fachbuchverlag 1986
22. Elektron Crosslinking AB: Trocknen von Beschichtungen mittels Elektronenstrahlung.pdf. Juni 2006. http://www.Crosslinking.com (Letzter Zugriff 28.12.08)
23. Enka, Firmenmaterial (1987): Beschichtung allgemein. Wuppertal 1987
24. Felger, H. K. (Hrsg.) (1986): Polyvinylchlorid. 2. Aufl. Bd. 2/2. Munich: Hanser 1986
25. Finke, W. (1994): Walzenberechnungen. Wiesbaden: Vieweg 1994
26. Finkmann, H. (1986): Herstellung von PVC-Compounds. Munich: Hanser 1986
27. Fitzgerald, R. (1991): Investitionsrechnung. Munich: Hanser 1991
28. Flury, F. u. a. (1962): Schädliche Gase. Berlin: Springer 1969
29. Fries, D. (1992): Textilveredlung Beschichtung. Ausbildungsmittel-Unter-richtshilfen. Eschborn 1992
30. Fritz, W. (1990): Reinigung von Abgasen. Würzburg: Vogel Verlag 1990

A. Giessmann, *Coating Substrates and Textiles,*
DOI 10.1007/978-3-642-29160-9, © Springer-Verlag Berlin Heidelberg 2012

31. Fuest, K. (2000): Elektrische Maschinen und Antriebe. 5. Aufl. Wiesbaden: Vieweg 2000
32. Gärweiler, A. (1989): Unternehmensstrategien. Frankfurt: Campus 1989
33. Geyer, E. (1987): Die Kreativität der Produktplanung. Landsberg: Moderne In-dustrie 1987
34. Giessmann, A. (1994): Anforderung und Auswahl eines Abluft-Reinigungs-systeme für die Beschichtungsindustrie. St. Gallen: Coating Verlag 1994. S. 222–225
35. Giessmann, H. (1982): Beschichten und Kaschieren von Geweben. Melliand Textil-berichte(1982) 11, S. 1348–1355
36. Hautz, E. (1992): Abgestimmter Informations- und Materialfluß über die Un-ternehmensgrenzen hinaus. Industrial Engineering (1992) 3, S. 4–7
37. Heeg, F.-J. (1991): Moderne Arbeitsorganisation. 2. Aufl. Munich: Hanser 1991
38. Heinrich, G. (1987): Energieökonomische Entstaubungsanlagen. Berlin: VEB Verlag Technik 1987
39. Henselder, R. (1990): TA Luft. Köln Bundesanzeiger Verlagsgesellschaft 1990
40. Hindemann, A. (1985): Zur Egalität beim diskontinuierlichen Färben. Informationsmaterial der Fa. Elf Atochem, Paris 1985
41. Hinterhuber, H. (1996): Strategische Unternehmensführung. 6. Aufl. Berlin: de Gruyter 1996
42. Johannaber, F. (1992): Kunststoffmaschinenführer. 3. Aufl. Munich: Hanser 1992
43. KEU Firmenprospekt (1989): Combustor. Krefeld 1989
44. Kolar, J. (1990): Stickstoffoxide und Luftreinhaltung. Berlin, Heidelberg, New York: Springer 1990
45. Kolbusch, A. (2001): Neue Vertriebsstrategien für ein Geschäftsfeld eines KMU des Maschinenbaus—unter Beachtung der Herausforderungen im internationa-len Wettbewerb. Diplomarbeit. Mönchengladbach 2001
46. Kranich, S. (1985): Staubreinhaltung. Mainz: O. Krausskopf Verlag 1985
47. Krelus Firmenprospekt (1995): Infrarotstrahler als Trocknungsprozeß. Hirsch-thal 1995
48. Lamalle, J. (1991): Technologie beim Pastenauftrag. Symposium der Fa. Sol-vay, Brüssel 1991
49. Lange Ritter Firmenprospekt (2008): Prepreg. http://www.lange-ritter.de/pdf/Prepreg.pdf (Last visit 20 December 08)
50. Lindner, H. (2001): Physik für Ingenieure. 16. Aufl. Leipzig: Fachbuchverlag 2001
51. Matek, W. (2001): Maschinenelemente. 15. Aufl. Wiesbaden: Vieweg 2001
52. Mecheels, St. (2001): Mit funktionellen und Intelligenten Bekleidungstextilien zu neuen Märkten. Vortag beim Forschungskuratorium 2001. Berlin 2001
53. Menig, H. (1977): Luftreinhaltung durch Adsorption, Absorption und Oxidation. Wiesbaden: Deutscher Fachzeitschriften Verlag 1977
54. Niemann Firmenprospekt (1991): Mischen und Dispergieren. Mannheim 1991
55. Offermann, P. (2000): Flächenbildungstechnik. Vorlesungsreihe „Verfahren und Maschinen der Textil- und Konfektionstechnik". TU Dresden 2000
56. Olfert, K. (1988): Investition. 4. Aufl. Ludwigshafen: Kiehler Verlag 1988
57. Peavy, H. (1985): Environmental Engineering. St. Louis: McGraw-Hill Book Company 1985
58. Plank, H. (2001): Einsatz von Textilien in Life Science. Vorlesungsreihe und Vortrag beim Forschungskuratorium 2001. Stuttgart 2001
59. v. Pupka, H. (1983): Investitionsplanung. Wiesbaden: Gabler 1983
60. Rammeisberg, R. (1992): Ist Logistik erlernbar? REFA Nachrichten (1992) l, S. 14–22
61. Reetz, E. (2001): Einführung in die Textilbeschichtung. Vortragsreihe beim 2. Coatema Coating Symposium, Dormagen 2001
62. REFA (Hrsg.) (1985): Methodenlehre der Planung und Steuerung. Bd. 4. Munich: Hanser 1985
63. REFA (Hrsg.) (1992): Informationsmanagement. Vorlesungsreihe zur Ausbildung zum REFA Ingenieur. Darmstadt 1992, S. 37–43
64. REFA (Hrsg.) (1994): Planung und Steuerung. Munich: Hanser 1994
65. Rigby, D. (1997): The world technical textile industry and its markets: Prospects to 2005. David Rigby Associates, Manchester 1997
66. Ringes, G. (1991): Produktinnovation. Berlin: Schmidt Verlag 1991

67. Rouette, H. K. (2001): Grundlagen der Textilveredlung. 14. Aufl. Frankfurt: Deutscher Fach verlag 2001
68. Samson Firmenprospekt (1994): Meß- und Regeleinrichtungen. Frankfurt 1994
69. Sattler, K. (1982): Umweltschutz Entsorgungstechnik. Würzburg: Vogel Verlag 1982
70. Schürmann, H. (2008): Konstruieren mit Faser-kunststoff-verbunden. 2nd edition Springer Verlag 2008
71. Schmidt, P. (1967): Ausführungsformen des Streichverfahrens. Schulungsunterlagen der Fa. Bayer Leverkusen 1967, S. 9–12
72. Sen, A. K. (1999): Coated Textiles. Ancaster: Technomics Publishing Co. 1999
73. Singer, U. (1991): Neues Verfahren für die Wirtschaftlichkeitsbeurteilung von Investitionen in neue Produktionstechnologien. Kostenrechnungspraxis (1991) 4, S. 63–70
74. Smith, W. C. (1999): Coated and laminated fabrics putting the industry in perspective. Industrial Textile Associates, Greer 1999
75. Spitzner, K. (1980): Textildruck. 3. Aufl. Leipzig: VFB Fachbuchverlag 1980
76. Tritron: http://www.tritron.Org/UV-Haertung-UV-Cure.12.0.html (Zugriff Dez. 2008)
77. Troge, A. (1985): Technik und Umwelt. Köln: Deutscher Instituts-Verlag 1985
78. Ungricht Firmenprospekt (1992): Gravurtechnologie. Mönchengladbach 1992
79. Ungricht Firmenprospekt (1994): Bonding Stars. Mönchengladbach 1994
80. Unterseh, H. (1993): Quo vadis Textilindustrie—gemeinsam die Zukunft meistern. Chemiefasern/Textilindustrie (1993) 11, S. 730–732
81. Vansteenkiste, P. (1978): Chemie-Technik. Stuttgart: Wissenschaftliche Verlagsgesellschaft 1978
82. VDI-Richtlinie 2280: Düsseldorf: VDI Verlag 1977
83. VDI-Richtlinie 3477: Düsseldorf: VDI Verlag 1984
84. VDI-Richtlinie 3478: Düsseldorf: VDI Verlag 1985
85. Wannagat, H. (1995): Walzenoberflächen. Gespräch mit dem Geschäftsführer der Fa. Wannagat. Witten, Februar 1993
86. Wehlmann, J. (1981): Kunstledertechnik. Leipzig: VFB Fachbuchverlag 1981
87. Weinhold, G. (1987): PVC-Additive zur Behandlung von Kunstleder. Schulungsunterlagen der Fa. Wacker. Burghausen 1987
88. Werner, A. (1986): Encyclopedia of PVC. New York: Litton Educational Publishing 1986
89. Werner, Firmenprospekt (1988): Misch- und Rührwerke. Mannheim 1988
90. Wikipedia, Braunsche Röhre, http://de.wikipedia.org/wiki/Braunsche_Röhre (Last visit 19 January 09)
91. Wikipedia, Faserverbundwerkstoff, http://de.wikipedia.org/wiki/Faserverbundwerkstoff (Last visit 20 December 08)
92. Wikipedia, Prepreg, http://de.wikipedia.org/wiki/Prepreg (Last visit 21 December 08)
93. Wikipedia (1): www.wikipedia.org, freely adapted from English text, keyword "Corona" (Last visit 02 January 09)
94. Wikipedia (2): www.wikipedia.org , keyword "Sol-Gel", (Last visit 03 January 09)
95. Wöhe, G. (1990): Einführung in die Allgemeine Betriebswirtschaftslehre. 17. Aufl. Munich: Vahlen Verlag 1990
96. Zuendel und Partner (1992): Kettenglieder eines Betriebes. Projektvortrag, Nettetal 1992

Index

A
Absorption process, 202
Accumulation trough, 56
Acrylate, 93, 109
Additional unit, 11
Adhesive, 42
Adhesive coat, 78
Adsorption process, 202, 204
Aerosols, 195
Agitator, 93
Air circulation tunnel, 62
Air contamination, 194
Air jet, 43
Air jet tunnel, 62, 64
Air knife, 29, 30
Air pollution, 195
Amortisation, 227
Amortisation period, 236
Annuity, 228
Application device, 27
Application precision, 75
Application roller, 42
Application system, 1, 16
Apply, 1

B
Base material, 102, 107
Batch winder, 53
Binding, 1
Biological sorption process, 205
Blow room, 104
Business strategy, 220

C
Calender, 51
Calendering, 51
Calendering process, 51
Card room, 105
Case knife, 29, 35

Catalytic converter, 207
Chemical fibres, 102
Clothing items, 5
Clothing materials, 116
Coagulation, 52
Coagulation process, 2
Coat, 2
Coat thickness, 83
Coating, 1, 5
Coating aggregate, 23
Coating industry, 101
Coating method, 78
Coating process, 1
Coating system, 12, 23
Coating weight, 75
Combustion process, 202, 207
Compensator, 56
Condensation process, 202
Contaminants, 207
Continuous systems, 16
Cooling, 71
Cooling rollers, 71
Corona pretreatment, 101
Corporate philosophy, 220
Cost elements, 222, 229
Cost factor, 219
Crush cut, 59
Cutting device, 59

D
Data, 76
Desk coater, 72
Dip roller, 42
Dipping, 49
Direct current motor, 58, 59
Direct process, 2
Discontinuous devices, 16
Dispersions, 27
Doctor knife form, 30

A. Giessmann, *Coating Substrates and Textiles,*
DOI 10.1007/978-3-642-29160-9, © Springer-Verlag Berlin Heidelberg 2012

Drive capacity, 58
Drive technology, 58
Drying, 61
Dust, 195
Dye, 48, 66

E
Embossing, 11, 77, 81
Emission, 196
Emission control, 198
End product, 2, 101
Energy costs, 224
Environmental protection, 101
Ex-protected, 110
Exhaust air, 198
Exhaust gases, 193
Explosion protection, 12
Extenders, 93, 95

F
Fabric, 32, 78, 105
Factory, 12
Fibrous material, 102
Fibrous web, 109
Filament, 102
Filter, 199
Final assembly, 77, 88
Finish, 11
Flexibility, 101, 229
Flock coating, 84
Floors, 113
Flow ratio, 74
Foam coating, 50
Foaming, 50
Foaming agents, 50
Foils, 1
Foulard, 49

G
Gap, 41
Gap setting, 32
Gases, 194
Gelling, 62
Glue, 1, 85
Grain diameter, 195
Grey products, 19

H
Heat radiation tunnel, 62
Heated roller, 65
Hot melt, 41

I
Imitation leather, 85, 109
Immissions, 196

Impregnating, 49
Infrared, 63
Inorganic fibres, 102
Inspect, 77, 88
Intelligent textiles, 3
Interest, 228
Interlaced fabric, 78, 105
Investment, 219
Investment spending, 223

K
Knife coating machine, 30
Knit fabrics, 105
Knitted fabric, 105

L
Laboratory, 15, 20
Laboratory equipment, 12
Lacquering, 81
Laminating, 1, 42, 85
Large batch winder, 89
Load cells, 73
Logistical integration, 19
Logistics chain, 17
Logistics integration, 12
Loops, 107

M
Maintenance costs, 223
Master batches (colours), 93
Material reserves, 56
Material tension, 43
Material transport, 53
Mayer bar system, 35
Microwaves, 76
Minimum exhaust volume flow, 70
Mixer, 52
Mixture, 214

N
Natural fibres, 102
Non-woven fabrics, 4
Non-woven material, 78

O
Odorants, 195
Operating costs, 223
Operating life, 227
Organosol, 1
Oxidation, 209

P
Packaging, 89
Paper, 1

Paste, 1
Performance elements, 219, 229
Personnel costs, 224
Pillow, 220
Pivot guides, 56
Plant layout, 11
Plasma pre-treatments, 101
Plastic, 45
Plasticiser, 81, 93
Plastisol, 1
Polyurethane, 93
Polyvinylchloride, 93
Powder, 41
Powder coating application, 42
Preparation, 11
Preparation of the pastes, 93
Print, 11, 40
Printing, 77, 80
Product strategy, 231
Production procedure, 15
Production process, 11
Production speed, 65
Profitability, 217
Profitability analysis, 219
PU, 109
PVC, 109

Q
Quality, 230
Quality monitoring, 75
Quantity, 229

R
Rain wear, 92
Raw material, 15
Recipe, 20
Reference recipe, 90
Release paper, 78
Repair costs, 225
Repayment, 228
Roller, 23
Roller gap, 83
Roller knife, 29
Rotation screen printing, 40
Rubber belt system, 29

S
Separator, 198, 199
Sheet, 1
Silhouette, 59
Silicon, 93, 109
Slot die nozzle application, 42
Smart textiles, 3
Smoke, 194
Solvent, 2, 12
Solvent mixer, 93

Soot, 194
Spinnable fibre, 102
Spinning, 105
Spiral knife, 35
Spraying, 46
Spread coating, 27
Spread coating method, 2
Substrate, 1
Substrate coating, 101
Support knife, 29
Surface, 38, 71, 107
Surface characteristic, 108
Surface effect, 82
Surface treatment, 79
Surface winder, 54
Suspension, 27

T
Table doctor knife, 29
Tarpaulin materials, 91
Tarpaulins, 114
Temperature, 64, 73
Temperature control unit, 70
Tensile force, 73
Tensile stress, 73
Tenter frames, 57
Textile coating, 2
Textile finishing, 2
Textiles, 1
Thermal oil, 66
Thermal oil boiler, 70
Thermal separation procedure, 204
Threads, 107
Three-phase drive, 58
Three-roller mill, 93, 95
Transfer coating, 78
Transfer process, 2
Transfer roller, 42
Tumbling, 82
Tunnel, 62, 85
Turret winders, 54

U
Units, 23

V
Vacuum filter, 93
Vacuum sinus dissolver, 93
Vapours, 194
Variability, 101
Viscosity, 28, 40, 81

W
Wall-mounted high-speed mixer, 93, 95
Web, 39, 55
Working width, 56

YOUR WORLD OF COATING

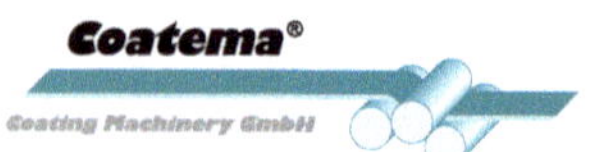
Coatema®
Coating Machinery GmbH

www.coatema.de